IMAGES
of America

DOWNTOWN GARY MILLRATS, POLITICS, AND US STEEL

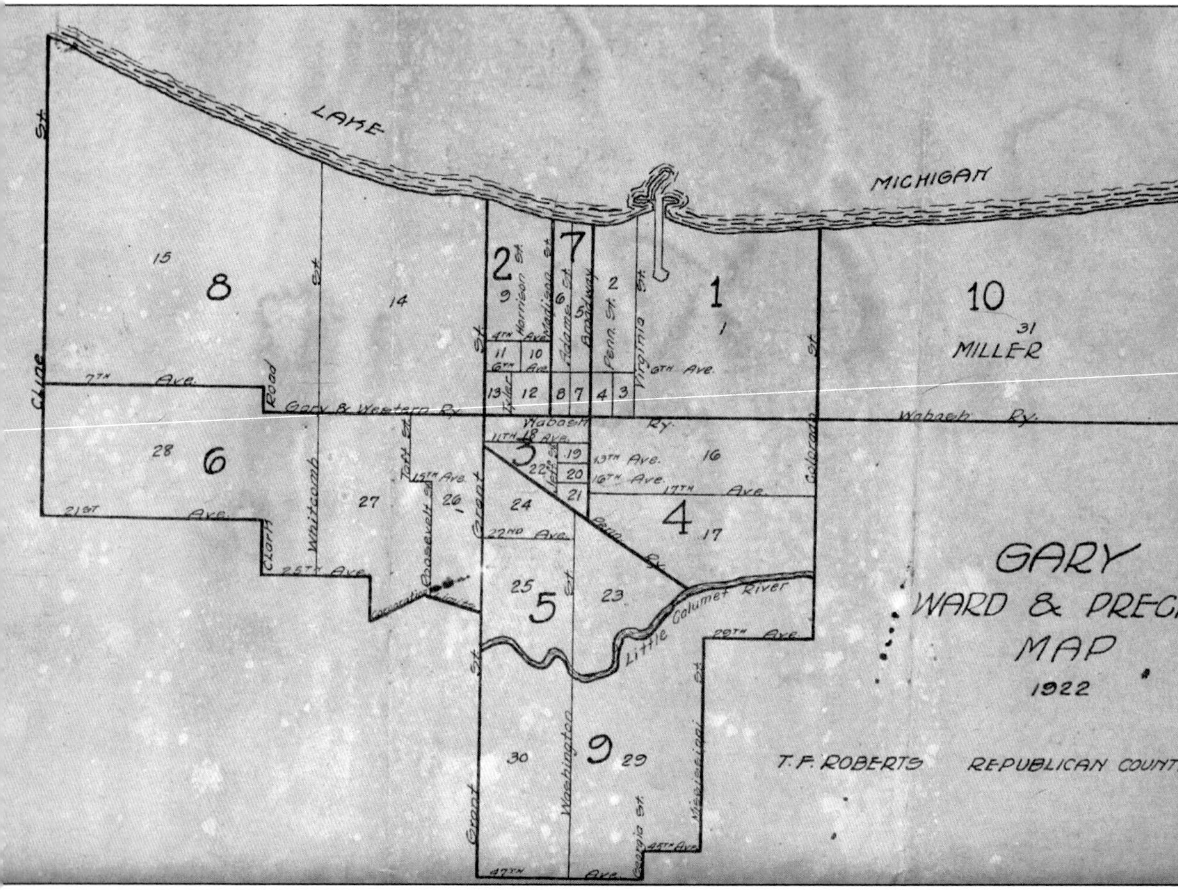

Map of Gary, Indiana. This 1922 district map of the city of Gary shows the boundaries as they would remain until the 1970s, when Black Oak was annexed. Downtown Gary and the adjacent Horace Mann and Emerson communities were located in Wards 2 and 7, and the western section of Ward 1. (Courtesy of Calumet Regional Archives.)

On the Cover: 25th Anniversary. In 1931, the city of Gary and the United States were in the midst of the Great Depression, yet there were things to cheer about in the Steel City. Gary was celebrating its Silver Jubilee. The city administration, US Steel, and the citizens all contributed to making it a memorable year. Pictured is the parade as it passed by the Lake Superior Court Building at Fourth Avenue and Broadway. The float was one of the many provided by US Steel's Gary Works. (Courtesy of Calumet Regional Archives.)

IMAGES
of America

Downtown Gary
Millrats, Politics, and US Steel

John C. Trafny and Diane F. Trafny

ARCADIA
PUBLISHING

ISBN 978-1-4671-0314-5

Published by Arcadia Publishing
Charleston, South Carolina

Printed in the United States of America

Library of Congress Control Number: 2018961612

For all general information, please contact Arcadia Publishing:
Telephone 843-853-2070
Fax 843-853-0044
E-mail sales@arcadiapublishing.com
For customer service and orders:
Toll-Free 1-888-313-2665

Visit us on the Internet at www.arcadiapublishing.com

To our parents, Steve and Wanda Trafny, who survived the Great Depression, lived through the horrors of World War II, and were a part of America's Greatest Generation.

Contents

Acknowledgments

Downtown Gary: Millrats, Politics, and US Steel would not have been possible without the support of a number of people. Some offered the use of treasured family photographs, while others took the time to give personal interviews. We were also provided technical help and assistance finding additional sources to use and research for our book. First, we want to offer a sincere thank-you to Steve McShane, the curator of the Calumet Regional Archives (CRA) at Indiana University Northwest in Gary. He provided great assistance in locating the photographs and sources that were needed for our project. In addition, he used a great deal of time to scan photographs and images for us.

Special thanks go to former east sider and graduate of St. Luke Parish School Collette Oram. She provided the beautiful wedding pictures of her parents for our use as well as photographs of first communions, graduations, and family events. Colette also provided background information on the parish, school, and former members of the old downtown parish, now Ss. Monica and Luke.

We appreciate the kind aid of Pamela Montgomery-Davis, secretary at Ss. Monica and Luke, for letting us go through the pictures the parish still has in its possession. It was also a wonderful experience to return to our former neighborhood parish after so many years away.

Thanks are also given to fellow alumni Judy and Dave Trathen, both from the Emerson class of 1968. They were able to locate and provide pictures of Emerson Elementary School when it was held in the middle school building in the late 1950s. We also appreciate the assistance of Michael Valtierra in allowing us to use personal family photographs and information on fellow Emerson graduates and Ss. Monica and Luke parishioners. A thank-you must also be given to Gary native Maurice Yancey, a graduate of Roosevelt High School and researcher, who provided background information on many of the old downtown stores.

We were able to obtain a great deal of background information on Gary's downtown, politics, and the steelworks from well-known local historians. Dr. James B. Lane's *The City of the Century* and his work with Dr. Ron Cohen, *Gary: A Pictorial History*, were great assets in finding exact dates, names, and places in the Steel City's past. Fellow Emerson alumnus Kendall F. Svengalis's *Gary: A Centennial Celebration* provided good background information on downtown churches.

We would also like to thank Thomas Campbell, former Emerson teacher, and Don Mueller, a graduate of Holy Angels School, for providing pictures. We are grateful for the late Dr. Lance Trusty, who helped in the creation of *The Polish Community of Gary* for Arcadia Publishing in 2000.

A sincere thank-you also goes out to Angel Hisnanick, our editor at Arcadia, for her help, guidance, and patience in getting our book completed.

Of course, no historical research can be perfect, and the authors assume full responsibility for any errors or omissions herein.

INTRODUCTION

Before the era of gigantic shopping malls, big-box stores, and online shopping, the commercial centers of major American cities were located in areas often referred to as downtown. In blue-collar industrial cities such as Gary, Indiana, downtown was the social, cultural, and political center of the community. Gary's downtown ran along both sides of Broadway, the Steel City's main artery, from Fourth Avenue to the Wabash Railroad tracks at Ninth Avenue.

From the 1920s through the 1960s, people from throughout the Calumet Region flocked to the Steel City's popular stores, theaters, and restaurants. If visitors were traveling from Chicago, Illinois, or from Michigan City or South Bend, Indiana, from the east, they took the Chicago, South Shore & South Bend Railroad. Those who lived in the nearby Emerson neighborhood to the east or the Horace Mann community to the west used the city's transit system, their personal car, or just walked the few blocks. It was the construction of the US Steel plant that brought about the birth of the City of the Century, Gary, Indiana, and its popular downtown.

In 1906, US Steel Corporation began construction of a modern steel plant on the southern shore of Lake Michigan, which, at the time, was a sparsely populated area of nothing more than sand dunes, swamps, and oak forests. The company selected that site as it was near major railroad and shipping lines and close to its South Works plant in South Chicago. It contained a large pool of workers available to build, and later operate, the new plant. Most of the workers were Eastern European immigrants. Others came from the rural South or the Appalachian coal towns of Pennsylvania and West Virginia. Nearly all were willing to work for whatever wages they were offered, and none of them, skilled or unskilled, were union men. Over the years, those who toiled in the mills became known as millrats.

When the first construction workers arrived to build the plant, most settled in tents and shanties near the main entrance to the mill at what is now Second Avenue and Broadway and just south of the Grand Calumet River. At first, many figured that the area was destined to become a company town similar to the coal and steel towns of Pennsylvania, but that did not happen as the company seemed more interested in its steel operations than city government. As more new residents settled in the city, neighborhood boundaries were established along man-made dividing lines, such as railroad tracks or major streets.

The Wabash Railroad line, near Ninth Avenue and Broadway, was the earliest line that divided the city, north and south, by ethnic group, race, and religion. Native born and better-educated Americans resided to the north of the tracks, while immigrants and unskilled workers lived on the south side. By the 1920s, downtown itself was divided by Broadway, the main north-south route. To the east was the Emerson neighborhood where most of the workers were skilled craftsmen, small business owners, and mostly blue-collar. The area west of Broadway was made up of mill supervisors, company executives, and professionals such as lawyers and doctors; however, some skilled craftsmen settled there as well. It became known as the Horace Mann neighborhood following construction of the new high school.

The downtown area was also home to many churches and synagogues. Like many urban areas, different ethnic and religious groups raised their own money to build their own houses of worship. One could find Polish, Russian, and Greek churches in Downtown Gary and surrounding communities. The Catholic churches of St. Luke on the city's east side and Holy Angels were built by the Irish. Other denominations that built beautiful houses of worship near Broadway were the Methodists, Lutherans, and Presbyterians. Two Jewish synagogues were located in the area.

From the time Gary was founded in 1906, politics played a significant part in the direction of the city concerning business and industrial expansion, residential development, and public school policies. Thomas Knotts, the city's first Democratic mayor, won in 1909 and fought the steel company over the direction of expansion as well as the Republican-controlled Lake County Council during his term. He was defeated in 1913, and the city was then run by the Republicans until 1935, when New Deal Democrats, under Pres. Franklin D. Roosevelt, began to take political control of the major urban areas of America. Since then, Gary has had only one Republican mayor.

Gary's politics, just like Lake County's, has had its share of scandals, indictments, and convictions over the years. Republican Roswell O. Johnson, the popular three-term mayor, was sent to federal prison for violating Prohibition in 1925. After his pardon by the president, he ran again in 1929 and won another term but was voted out during the Great Depression of the 1930s. George Chacharis, popular Democrat of the 1950s and early 1960s, also went to the federal lockup but never ran again. After his release, many voters said Cha-Cha, as he was called, knew how to run a city and that they would vote for him again. Even today, years after his death, many who were around in the 1950s and early 1960s have stories to share about the former mayor. Downtown Gary was also the place to see and sometimes meet presidential candidates during election years.

Gary may have been considered a gritty mill town by outsiders and visitors, but its downtown businesses still drew people from throughout the Calumet Region to shop and seek bargains, enjoy the food and environment of classic restaurants, and get tickets for the theater or latest movies. Before the age of television, families spent Saturdays downtown. Sunday shopping did not exist prior to the 1950s except for at corner drugstores or gas stations. Sometime during the late 1950s, stores opened on Sundays for business but only during the Christmas holidays.

Downtown Gary had a number of theaters that ran many of the classic blockbuster movies that drew packed houses each weekend. In addition, Saturday morning matinees drew loads of children. Opened on Broadway in 1925, the Palace was constructed in the style of a palace courtyard during a starry night. A few blocks away was the State, a more modern theater that was completed in 1937. The Gary Theater offered movies, vaudeville acts, live theater productions, and burlesque.

Shoppers were offered a host of stores in Downtown Gary. Large national chains like Sears, J.C. Penney, Florsheim, and S.S. Kresge Co., and Chicago-based stores like Goldblatt Bros. became popular with blue-collar families, especially those who wanted a good deal on furniture or appliances. H. Gordon & Sons, which opened on Broadway in the early 1920s, became one of the area's premier clothing stores. Others included Pearson, a women's clothing store, and Henry C. Lytton & Sons, menswear. Baby boomers may recall Comay's Jewelers with its record shop, Tom Olesker's, W.T. Grant, and Robert Hall Clothing on east Fifth Avenue. No matter the store, sales associates asked shoppers, "May I help you?"

Along Fifth Avenue visitors could patronize Olsen Cadillac, Baker Chevrolet, and Baruch Olds. Bakeries, such as Cake Box and Sno-White, provided delicious baked goods. Slicks Laundry, the Blackstone, the Lighthouse, Walts, and Gary Camera were other businesses located along that street. In addition, there were plenty of taverns in the area. They included Parkway, Cozy Corner, Trainor's, the Spitfire Lounge, the Ingot Inn, and a host of others. On payday Mondays, the saloons did good business as steelworkers cashed checks there instead of the banks. It was, after all, a steel town.

Sadly, by the 1960s, with the move to the suburbs and changing demographics, Gary's downtown went into a speedy decline. The growth of shopping centers and the new South Lake Mall in Hobart, Indiana, drew away business by the early 1970s. Downtown Gary never recovered. Politics, crime, and middle-class flight brought an end to a classic period in Gary's history that lasted nearly 50 years. Little except vacant buildings and empty lots remain there today. The people and businesses may be gone, but for many former residents and others who once shared enjoyable times there, the memories remain. It is for those who were part of that era and those who only heard about Downtown Gary that this book was written.

One

Millrat Neighborhoods

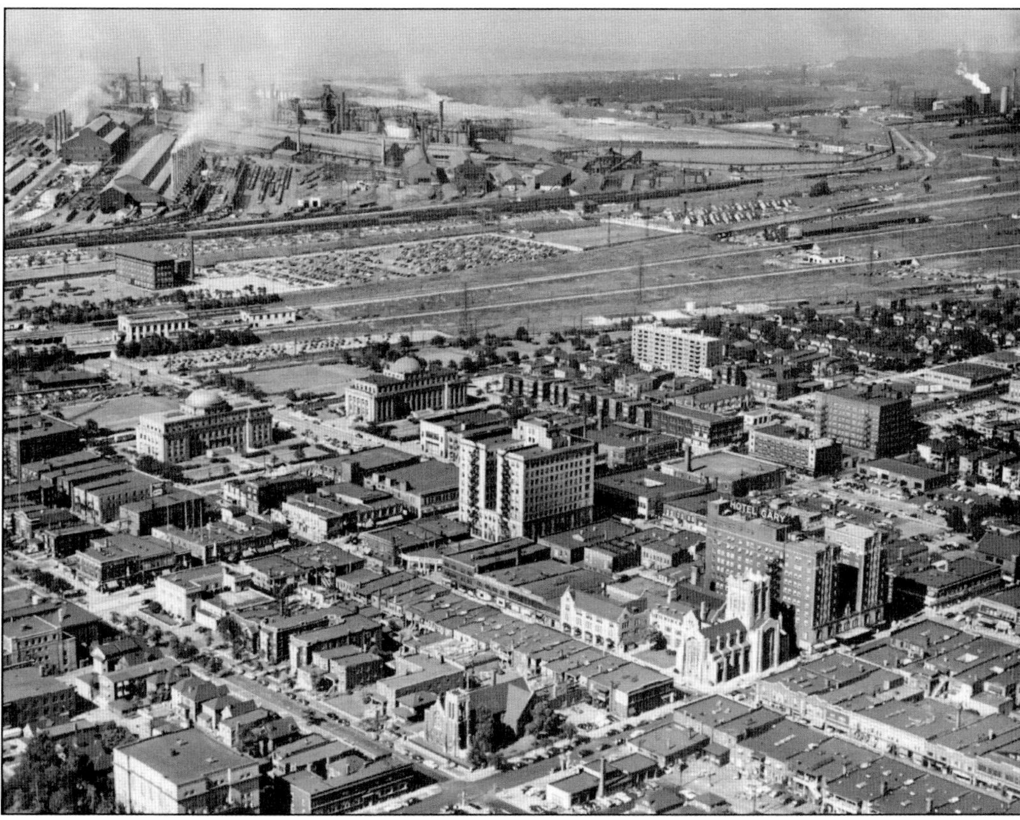

Downtown Gary, Aerial View, 1950. Long ago, before the growth of suburbia and indoor malls, it was the downtown section of urban America where business, politics, and commerce were located. Everything was just a trolley car, bus ride, or walk away, and that was the case for Downtown Gary, Indiana, the Steel City, from the 1920s through the 1960s. In this aerial shot, the steel mills can be seen in the background, while city hall and the Lake County Superior Court Building, both domed, are left of center. Broadway, Gary's main commercial strip, runs between them. To the east and west sides of Broadway are the Emerson and Horace Mann neighborhoods. Most of the residents there worked for US Steel; they were often called millrats. (Courtesy of CRA.)

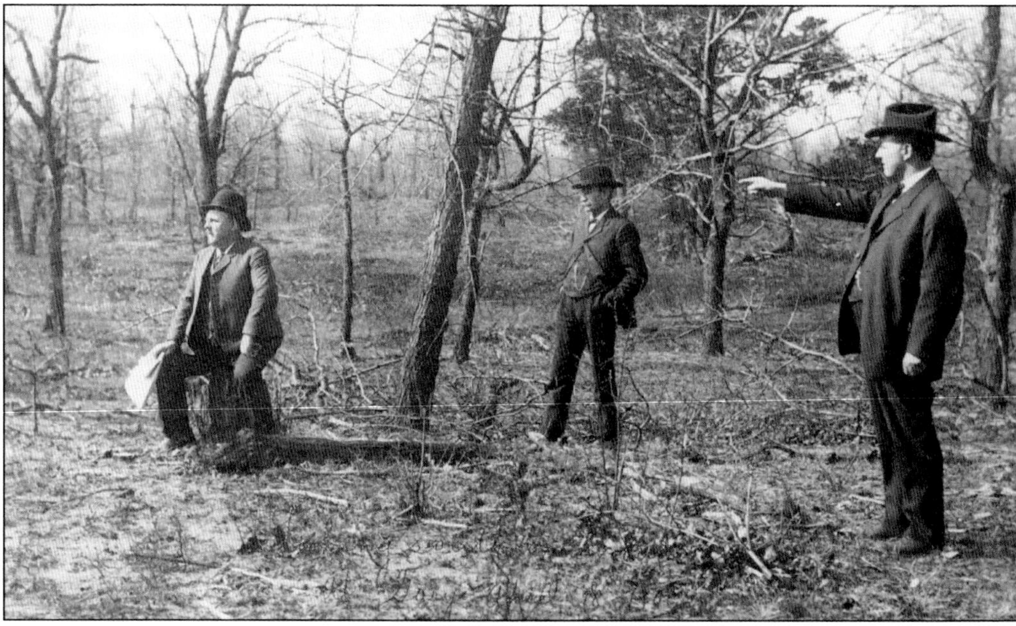

Birth of the Steel City, 1906. In March 1906, Thomas Knotts and his brother Armanis, center, marked the area that would soon become the intersection of Fifth Avenue and Broadway. Though the area was nothing but sand dunes, swamps, and vegetation, once construction began, a massive steel plant and modern industrial city would emerge. It became Gary, Indiana, the City of the Century. (Courtesy of CRA.)

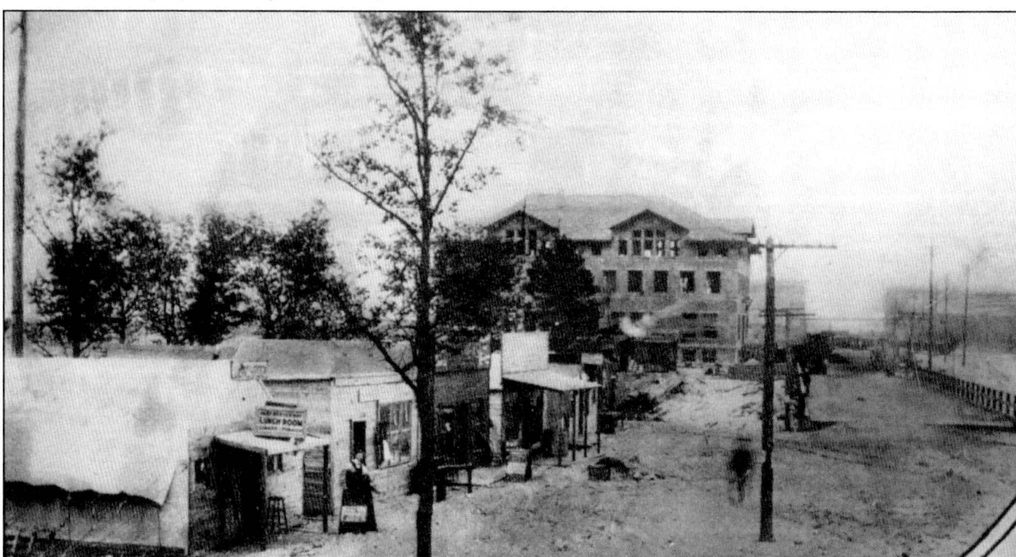

Shantytown, 1906. No living quarters were available for the first work crews that arrived to prepare the land for the construction of the mill and town. Tents were set up near the plant site, and soon, a shantytown evolved near the north bank of the Grand Calumet River. The first wooden structure was McFadden's Flats, named after Maj. John McFadden. The charge to stay was two bits (25¢) a night, or a buck a week. For the most part, mill officials stayed at the Calumet Gun Club's cottages. Pictured is a 1906 work site. (Courtesy of CRA.)

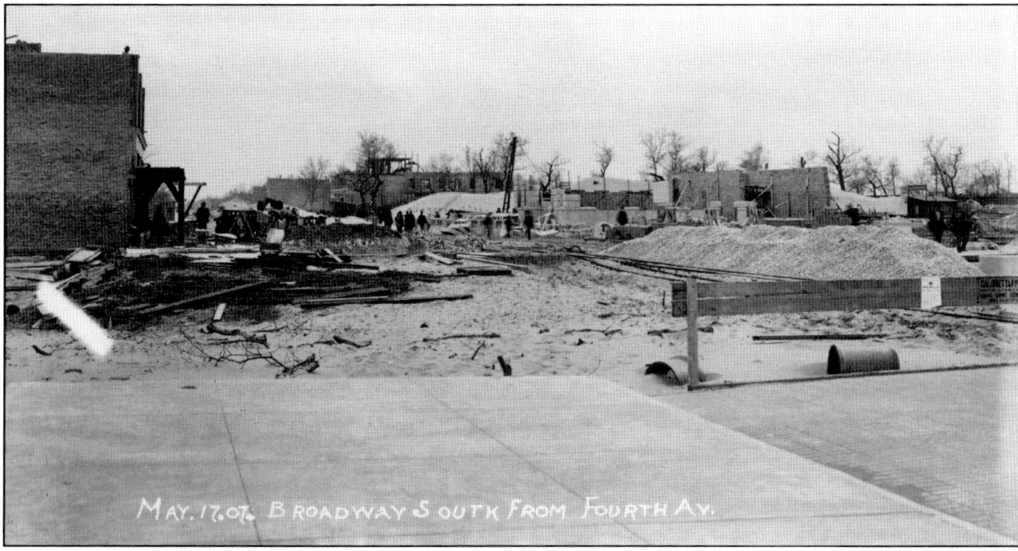

FOURTH AVENUE AND BROADWAY, 1907. South of the Grand Calumet River, construction work on Broadway's commercial district exploded. New structures went up for businesses that catered to the residents of the City of the Century. Pictured is the intersection of Fourth Avenue and Broadway. (Courtesy of CRA.)

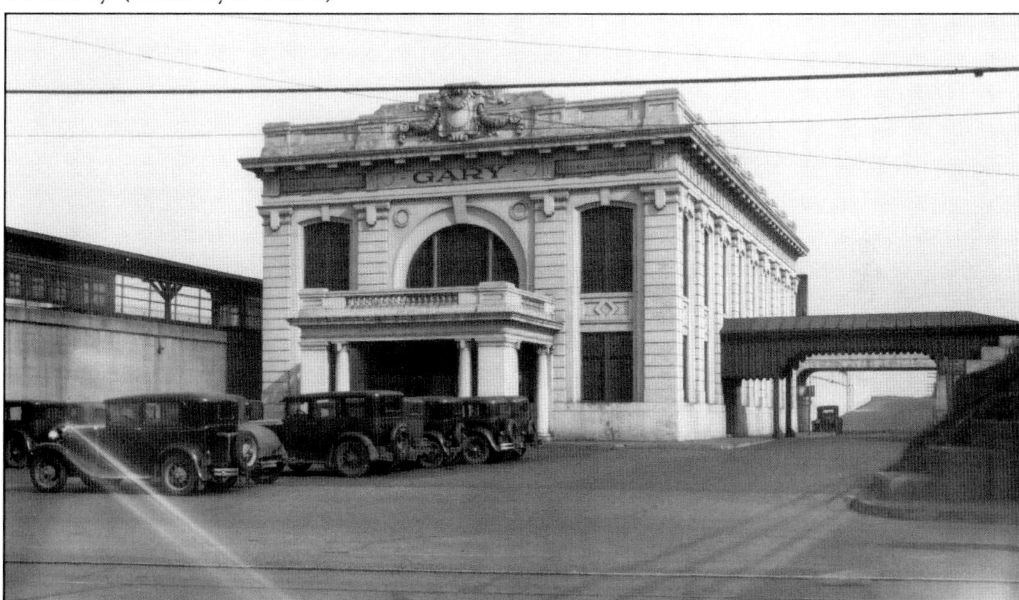

BALTIMORE & OHIO RAILROAD STATION, 1910. Railroads provided the best means for cross-country travel during the first half of the 20th century, and Gary was situated along many of the major east-west routes. The New York Central, Baltimore & Ohio (B&O), Pennsylvania, and Michigan Central railroads had lines that ran passenger service through different parts of the city. The New York Central and B&O station was located near Second Avenue and Broadway and operated from 1908 through the early 1970s, when it closed its doors for good. The station, pictured here in 1910, served passengers traveling east to New York, Baltimore, Washington, DC, and Pittsburgh. This image was part of the early postcard series that the City of Gary provided travelers and visitors to boost tourism from the early 1900s through the 1960s. (Courtesy of CRA.)

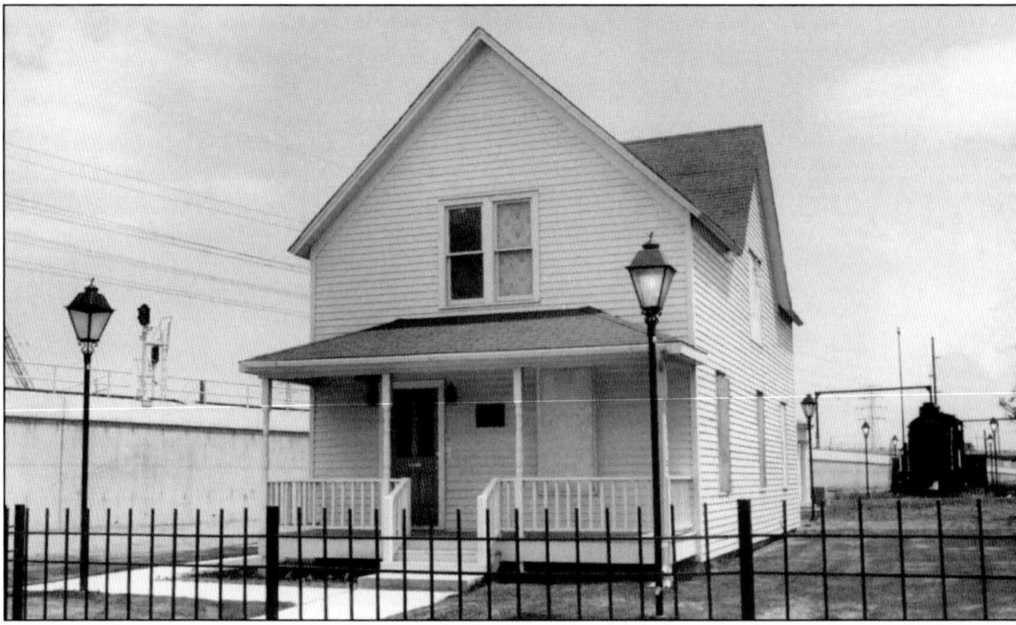

Gary Land Company. One of Gary's earliest structures was the Gary Land Company Building. Located along Broadway near Fourth Avenue, the two-story structure originally served as city hall. It once housed the Gary Historical Society and is now situated in Gateway Park along Fourth Avenue near the present-day city hall building. The Gary Land Company Building, as well as an old Elgin, Joliet & Eastern Railway locomotive, is pictured here in the 1980s. (Courtesy of CRA.)

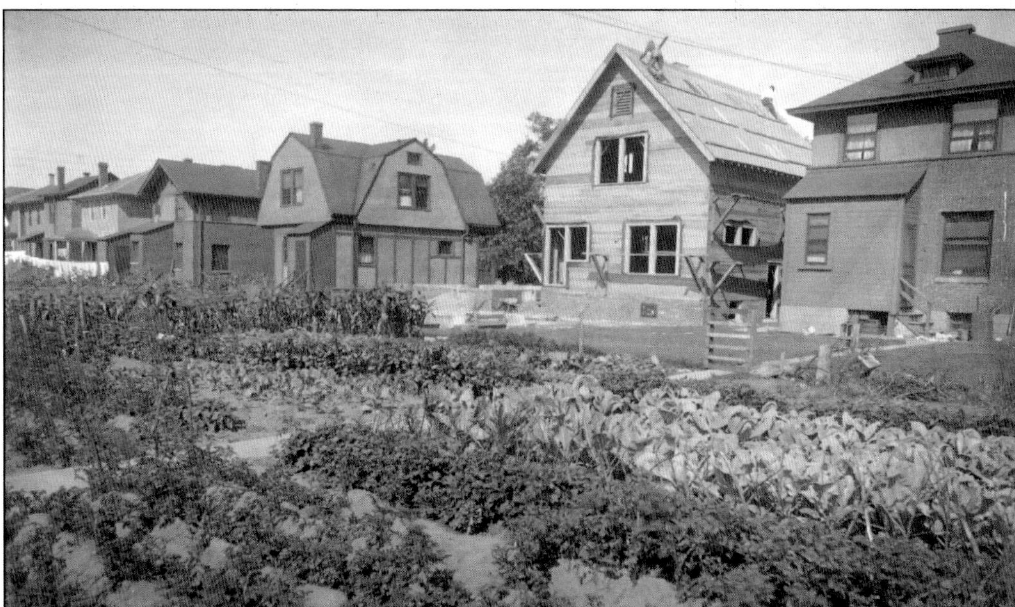

House at 552 Delaware, 1917. In addition to single-family houses and bungalows, many apartments and two-flats were constructed in Downtown Gary. The building at center had a basement with a laundry area and space for a furnace and coal storage. The main floor was the landlord's residence, while the second floor had rentals front and rear. A World War I victory garden is in the foreground. (Courtesy of CRA.)

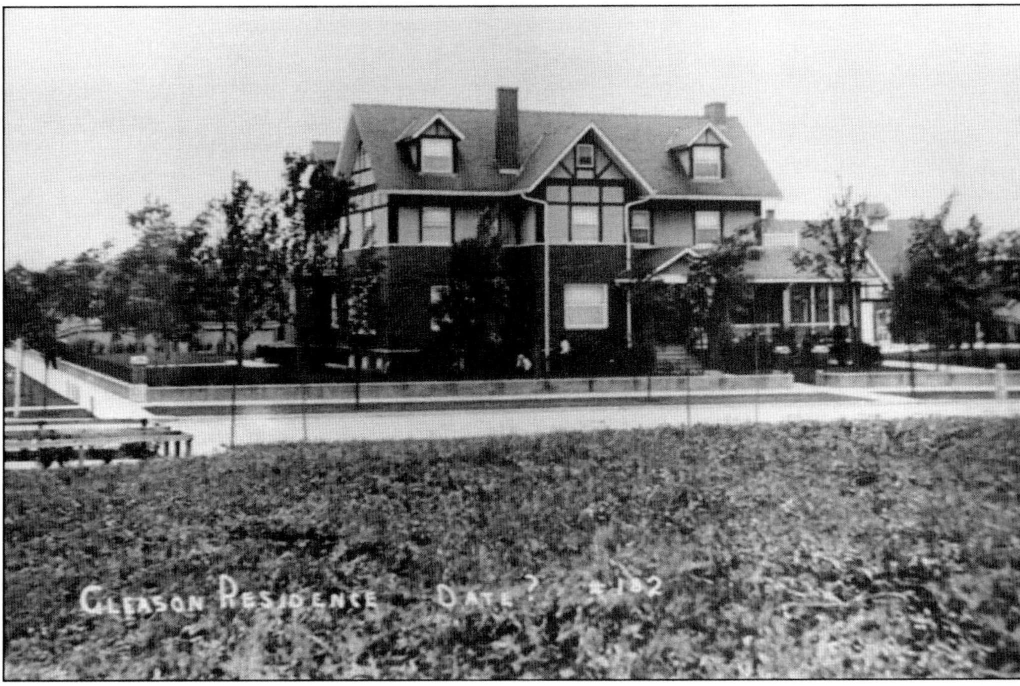

GLEASON HOME. Gary's west side was also an attractive neighborhood for doctors, lawyers, and executives. The Gary Works' first superintendent, William P. Gleason, resided in this elegant mansion. The building was later purchased by the Catholic Diocese of Gary. It is pictured here around 1917. (Courtesy of CRA.)

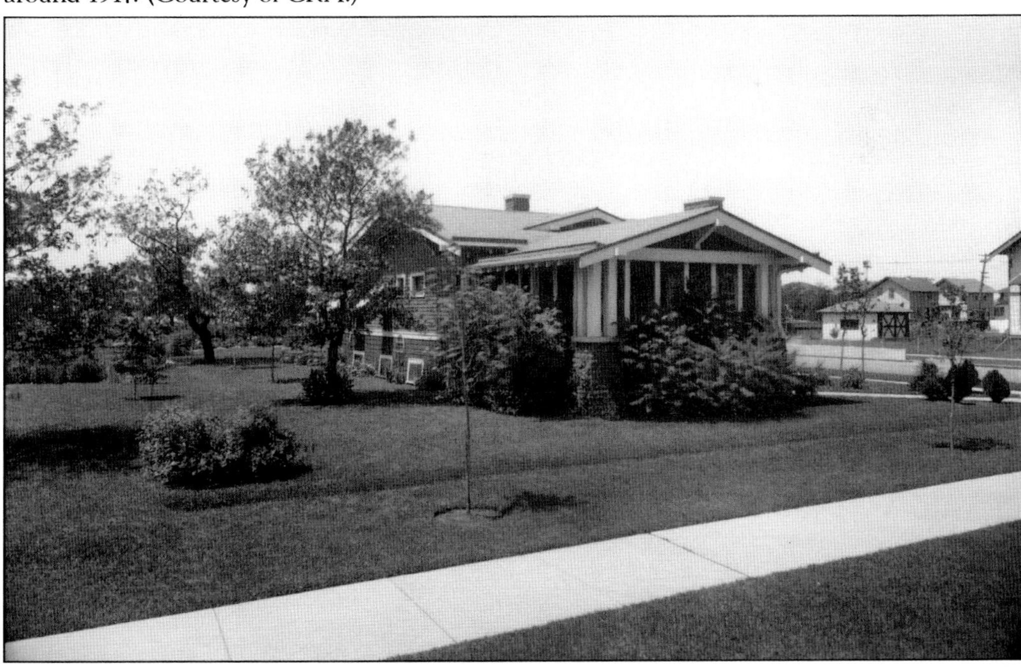

BUNGALOW HOUSE, 1915. As new residential housing went up across the city, one of the popular home styles was the single-family bungalow. Bungalows were ideal and less expensive for young families moving to the area. Pictured is a home at 704 Fillmore Street. (Courtesy of CRA.)

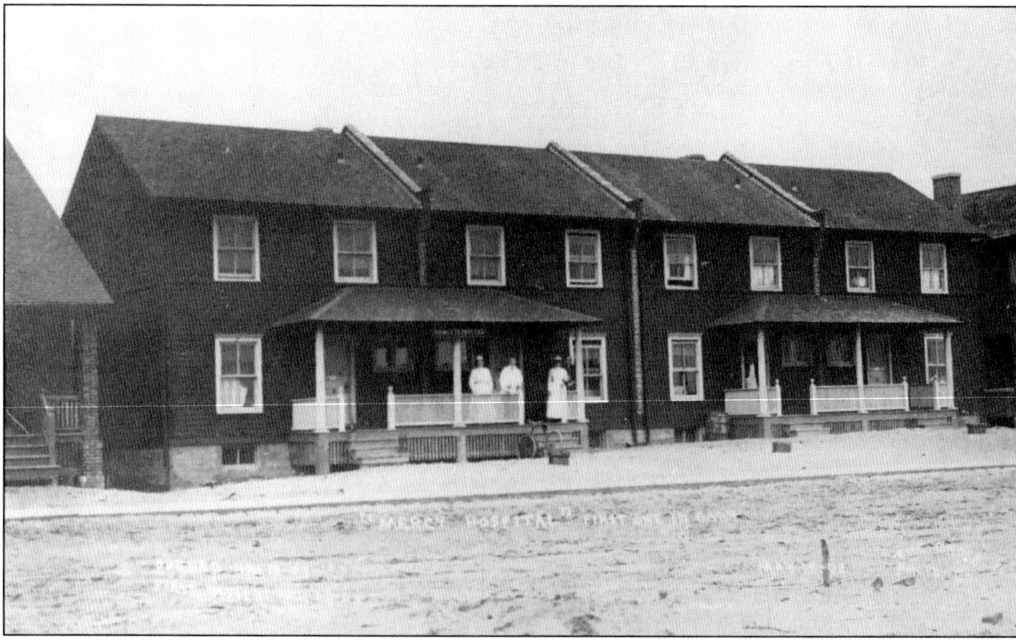

Mercy Hospital, 1908. Gary's first hospital, Mercy, opened its doors to the public in 1908. Located at Seventh Avenue and Carolina Street, across the street from Emerson School, the medical facility was run by the Roman Catholic Church. A few years later, the hospital moved operations to Sixth Avenue and Tyler Street, across from Holy Angels, until it closed in the 1990s. (Courtesy of CRA.)

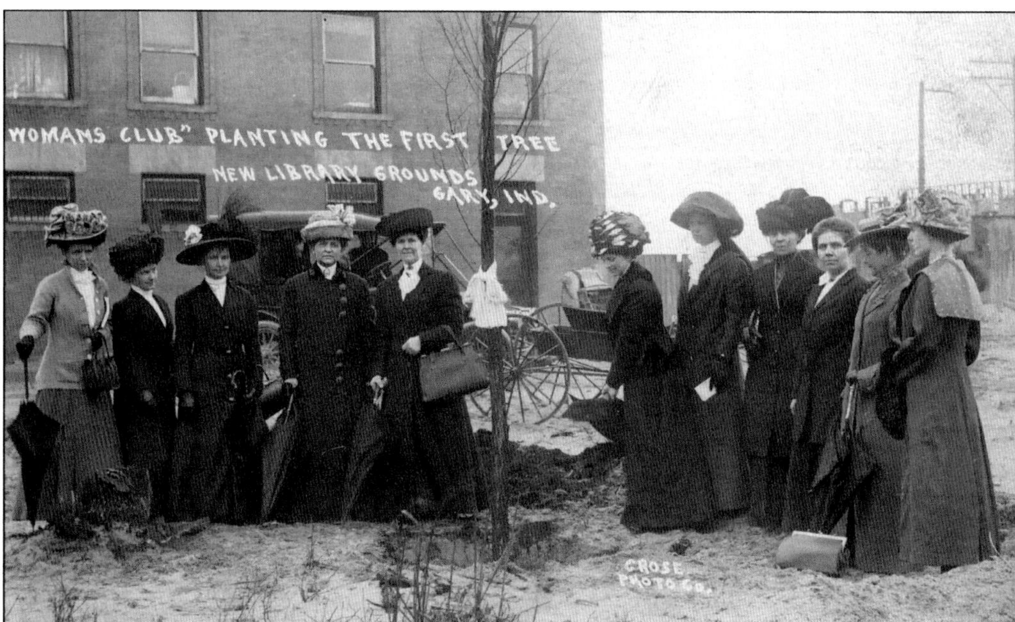

Ladies Planting Trees, 1912. As the young steel town grew, a number of community groups were organized to improve the growing neighborhoods of the city. Here, a west side ladies' society is seen plating trees near a library outside of Jefferson School about 1912. (Courtesy of CRA.)

Jefferson Park, 1910. West of Broadway residents could relax and enjoy themselves at Jefferson Park, located at Jefferson Street between Sixth and Seventh Avenues. This c. 1910 photograph shows the iconic water tower and pumping station. (Courtesy of CRA.)

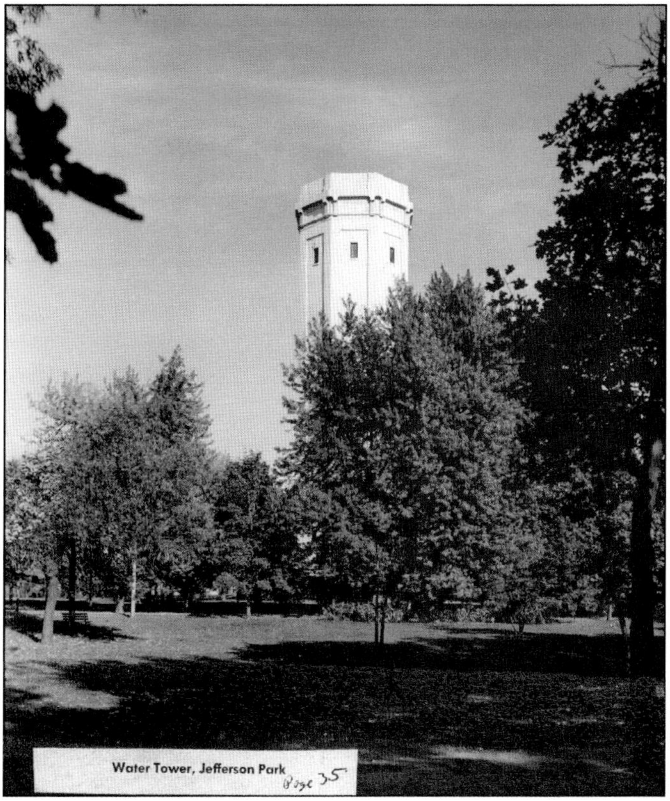

Post-Tribune, 1931. At one time, the *Post-Tribune*, Gary's newspaper, had its offices and printing operations in the 400 block of Broadway, seen here in early 1930s. Pedestrians along the east side of Broadway could watch the presses and printers working to produce the next edition. (Courtesy of CRA.)

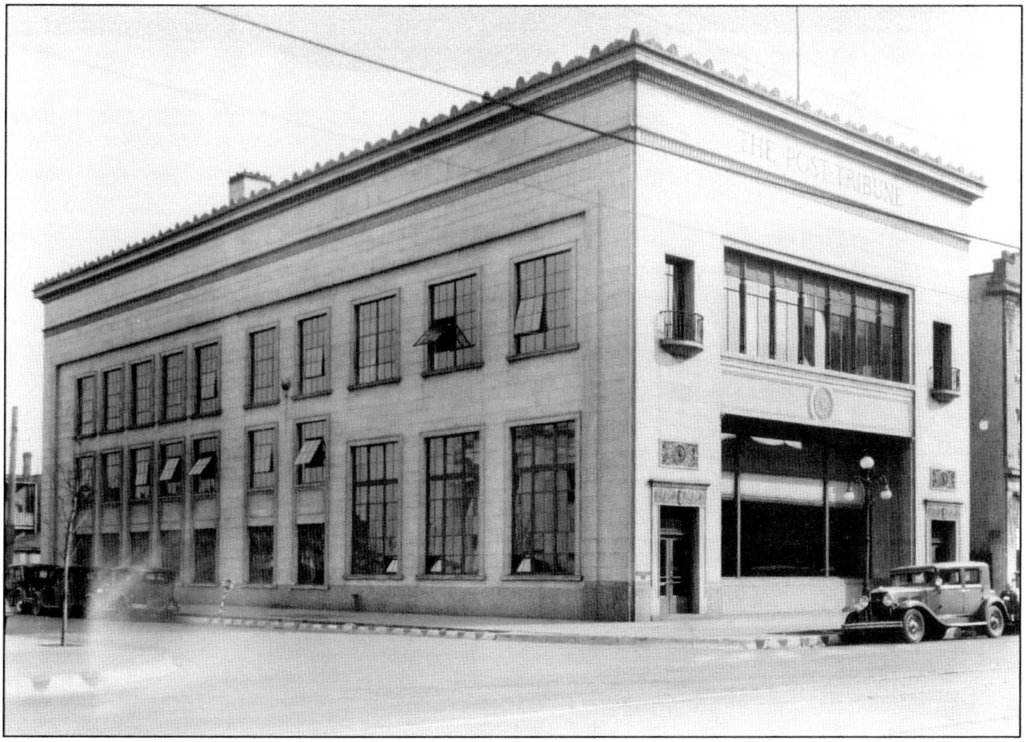

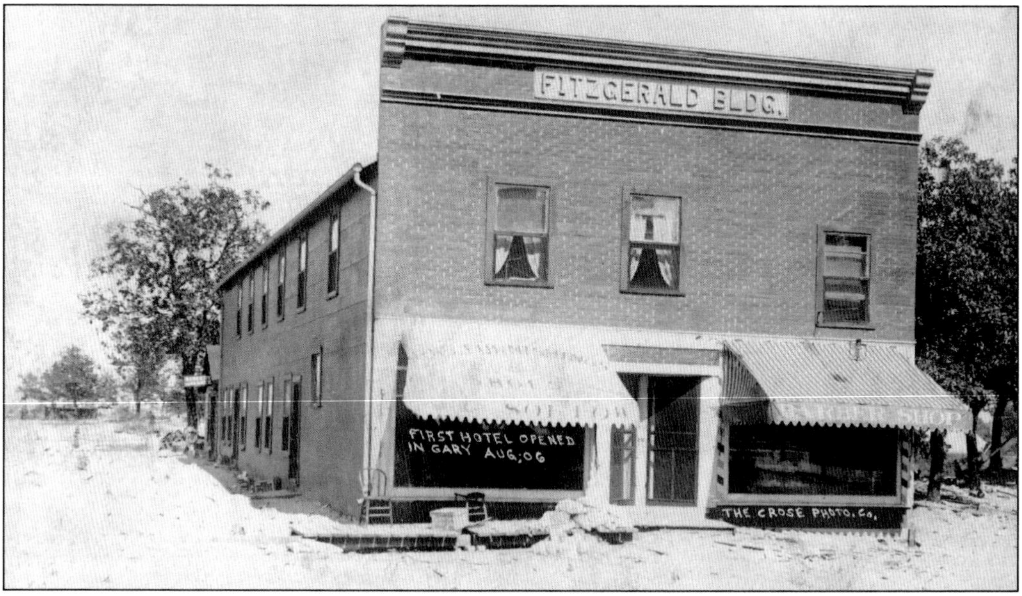

Fitzgerald Building, 1906. Gary's first hotel was constructed near Fourth Avenue and Connecticut Street close to US Steel's main gate at Broadway. Though not in the class of the Hotel Gary and Victoria Hotel that were built later, it provided customers a decent place to stay. The main floor had a saloon on the left and a barbershop on the right. After World War II, the new owners of the saloon renamed it the Spitfire Lounge in honor of the brave Royal Air Force (RAF) pilots who fought in the Battle of Britain in 1940. The owners were Americans who volunteered to serve in the RAF before the United States entered the war in 1941. (Courtesy of CRA.)

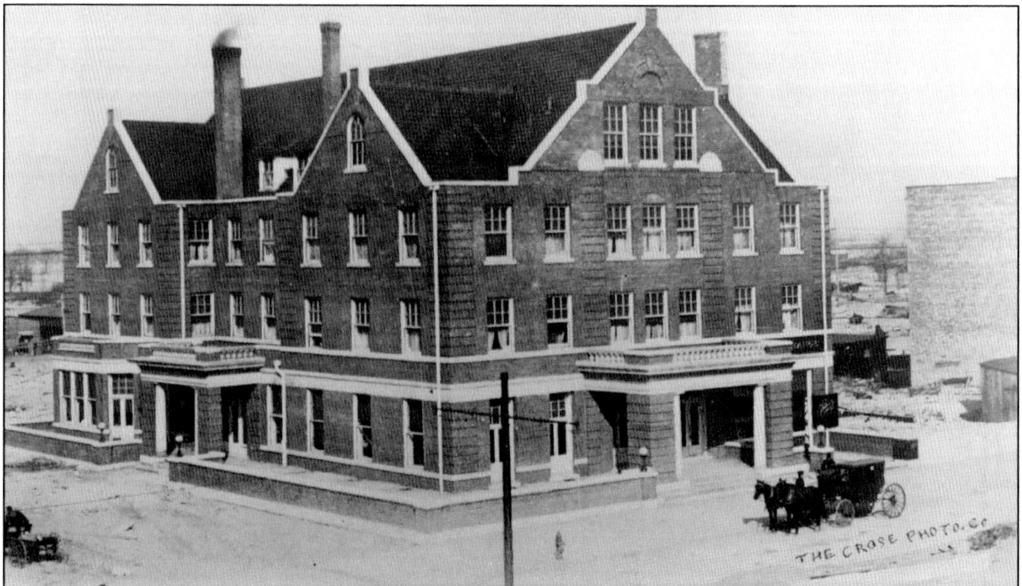

First Hotel Gary. The first Hotel Gary was constructed prior to the Great War of 1914–1918 at the corner of Sixth Avenue and Broadway to meet the increased needs of travelers, steel executives, and tourists. Gary's first hotel at that intersection was a wooden structure that was replaced by a modern brick and steel building during the boom of the mid-1920s. (Courtesy of CRA.)

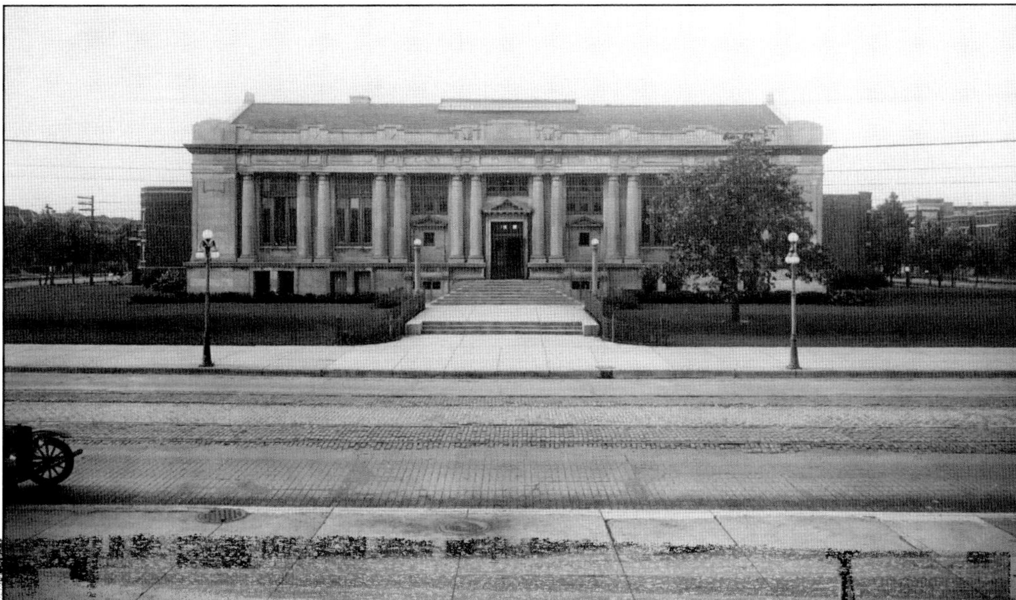

PUBLIC LIBRARY, 1913. Pictured is the Gary Public Library in 1913. The structure, located at Fifth Avenue and Adams Street, was situated on land donated through the efforts of Judge Elbert Gary and the Gary Land Company. The building contained a large collection of books, periodicals, and documents available for study. Librarians were always available for assistance and offered programs to students to provide help in their studies. (Courtesy of CRA.)

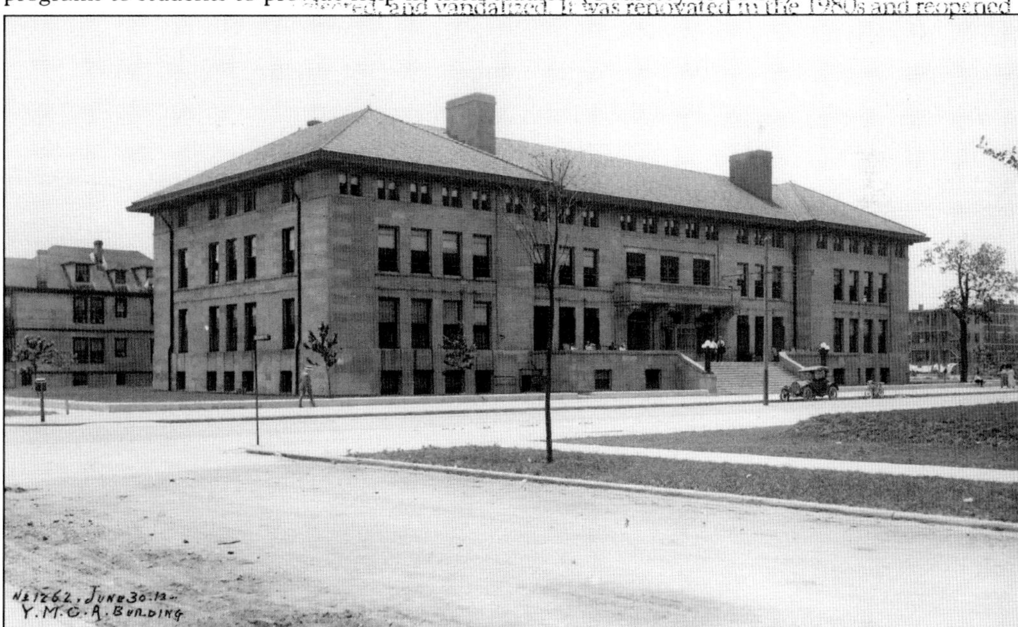

YMCA, 1910. Through the encouragement of Judge Elbert Gary, the Gary Land Company set aside property for charitable organizations, hospitals, and churches. The Young Men's Christian Association was given property at the southeast corner of Fifth Avenue and Jefferson Street to erect its building in 1910. A block east of Broadway on Sixth Avenue, the Young Women's Christian Association building was constructed a few years later. (Courtesy of CRA.)

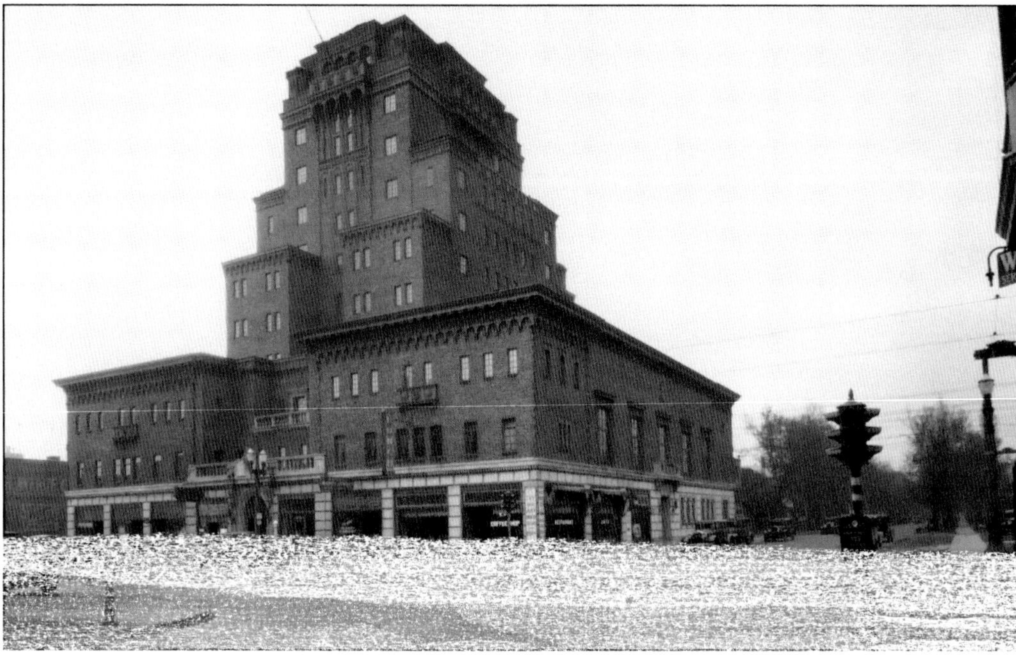

Knights of Columbus, 1931. Another of the grand buildings constructed during the boom of the 1920s was the Knights of Columbus building at Fifth Avenue and Madison Street. The 10-story structure contained shops, a gym, ballroom, restaurant, and hotel rooms. By the 1960s, the building was run down, neglected, and vandalized. It was renovated in the 1980s and reopened as senior housing. (Courtesy of CRA.)

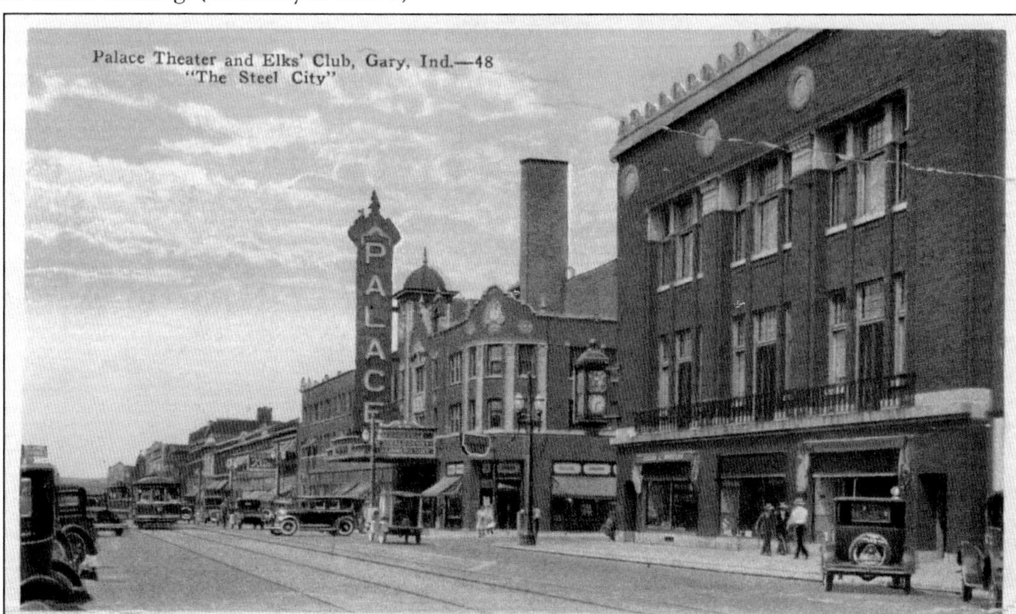

Palace Theater, 1925. In 1925, the Palace Theater opened its doors to show the latest in first-run Hollywood films. Located at Eighth Avenue and Broadway, the theater also held musical concerts, live acts, and some high school commencements. This postcard depicts the theater's facade. H. Gordon & Sons Department Store is at right. (Courtesy of CRA.)

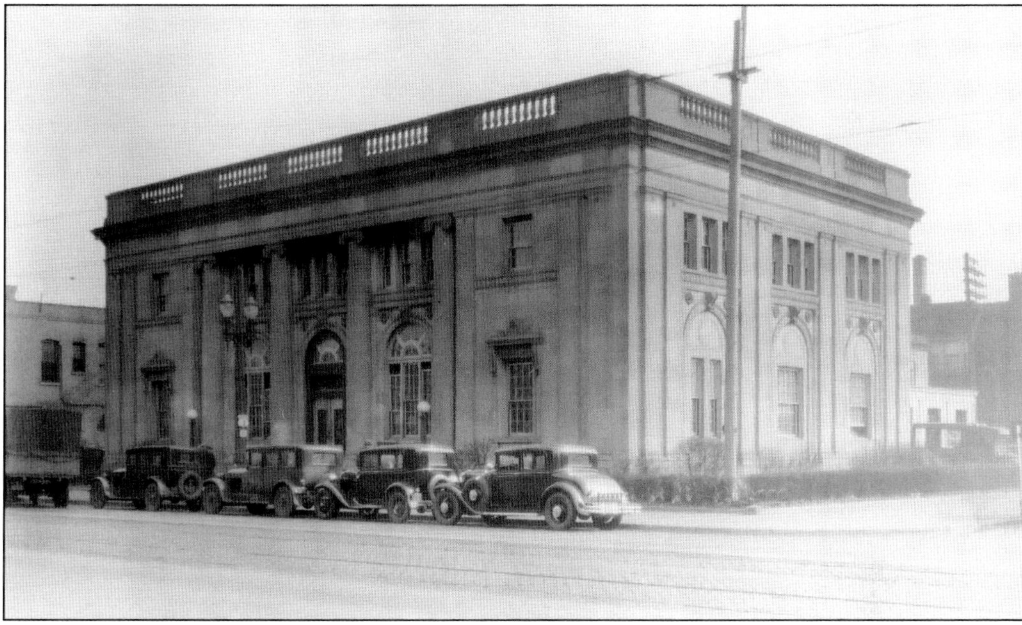

Old Main Post Office, 1931. Gary's main post office originally stood at the southeast corner of Fifth Avenue and Adams Street until the 1940s. The main branch then moved to Sixth Avenue and Massachusetts Street. It is pictured here in 1931. (Courtesy of CRA.)

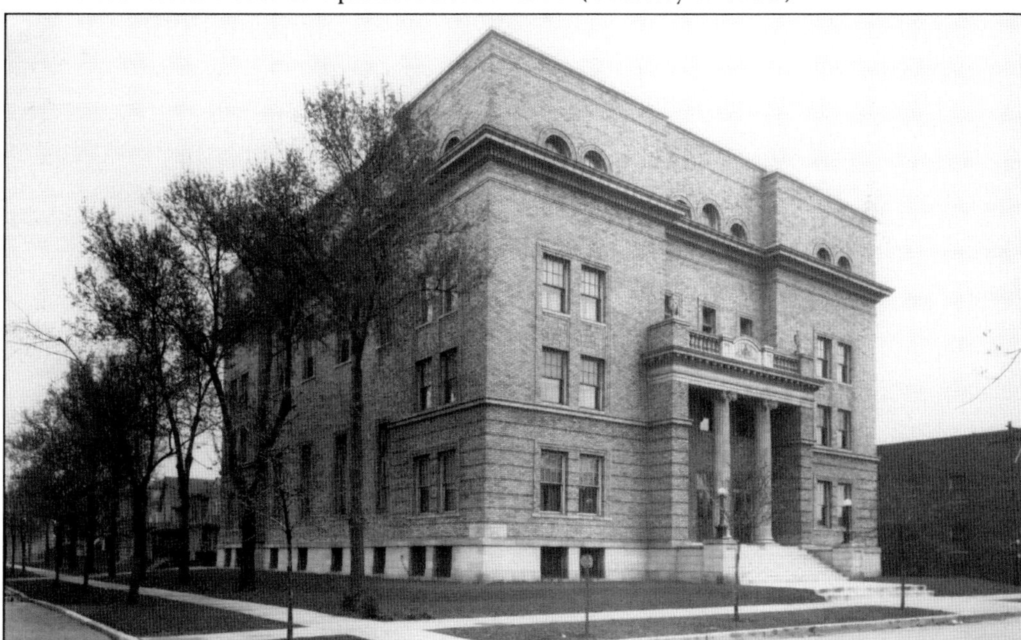

Masonic Temple, 1931. Members of Gary's Masonic lodge originally met in the old Hotel Gary at Sixth Avenue and Broadway. In the 1920s, members decided not to renew their lease and move to another location. They built a new structure at Sixth Avenue and Jefferson Street. The cornerstone was laid in 1925, and the building was completed in the late 1920s. It is pictured in 1931. (Courtesy of CRA.)

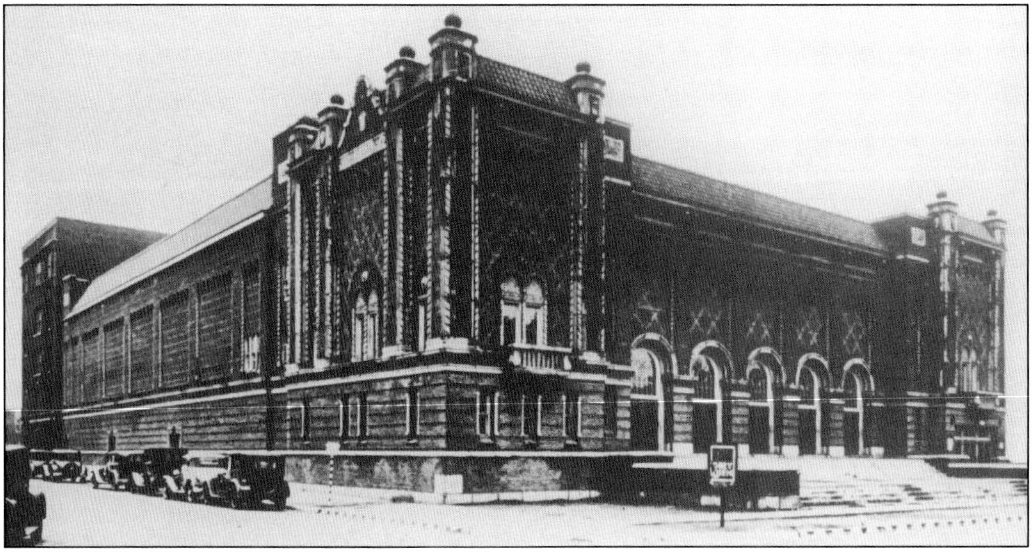

MEMORIAL AUDITORIUM, 1930. During the early 1920s, Downtown Gary experienced a building boom, which transformed the physical appearance of the Steel City. The boom continued until 1929 and resulted in the construction of the Hotel Gary, the Palace Theater, City Methodist Church, and the Memorial Civic Auditorium, pictured here. Built in 1927, the auditorium held concerts, plays, civic events, high school commencements, and basketball games. (Courtesy of CRA.)

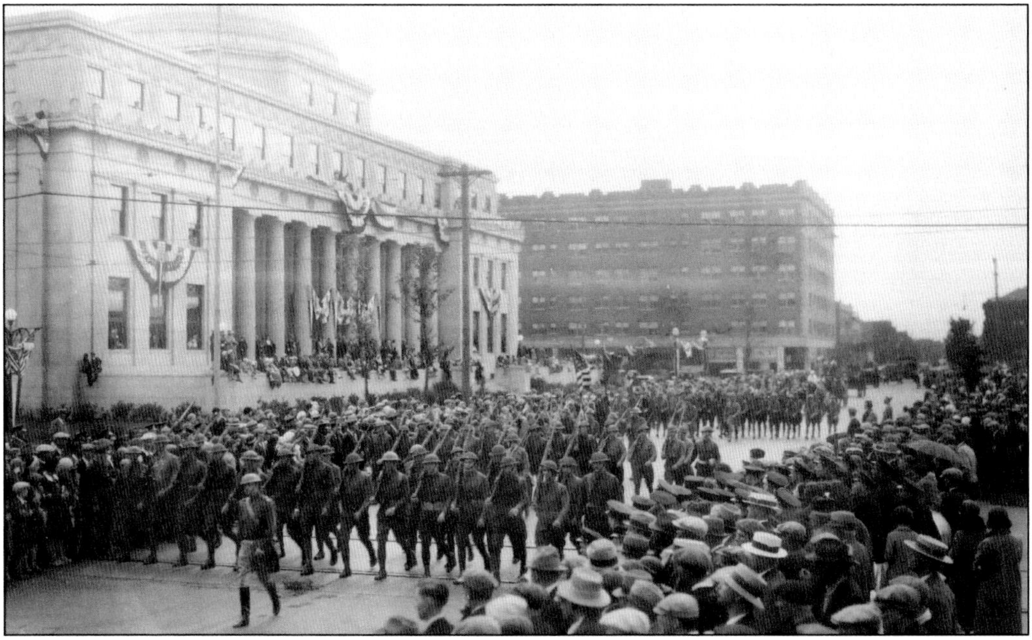

SILVER JUBILEE PARADE, 1931. Gary observed its Silver Jubilee in 1931, and a number of festivals, parades, and concerts were held throughout the city that summer. Politicians, prominent local people, and visitors gave speeches. Many historians consider 1931 the worst year of the Great Depression. For the citizens of the Steel City, the festive events took their minds off the stress and gloom of that economic disaster, at least for a few days. Pictured are members of the Indiana State Militia marching east on Fourth Avenue and Broadway. The Lake County Superior Court Building is seen at left. (Courtesy of CRA.)

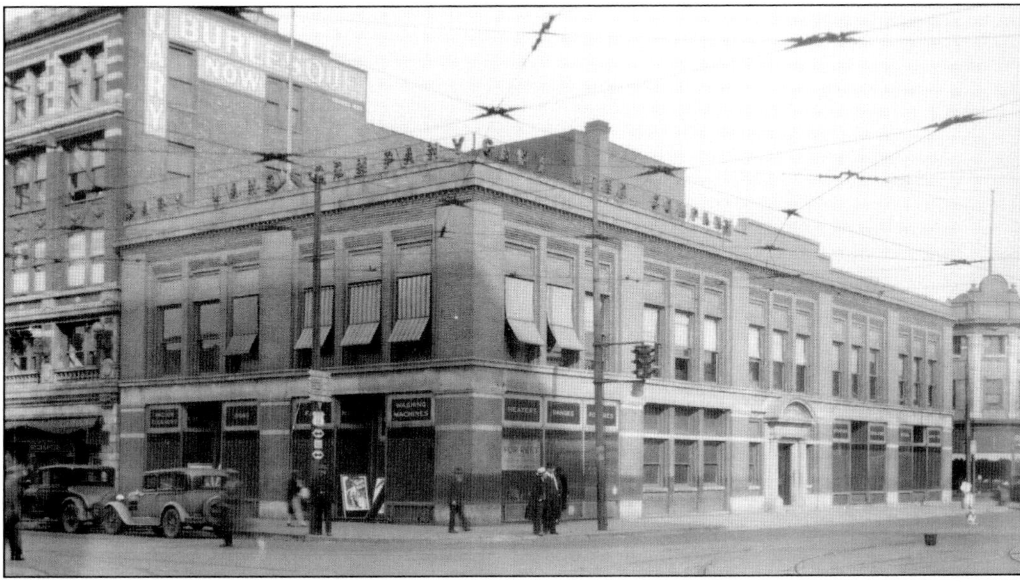

The Gary Theater. From the 1920s through the post–World War II years, the Gary Theater, located at the northeast corner of Fifth Avenue and Broadway, was an entertainment center in the Steel City. During the silent movie era, some patrons enjoyed adventure films featuring Douglas Fairbanks Sr., Rudolph Valentino, and Mary Pickford, while others watched the exploits of Tom Mix or William Hart in Westerns. In addition to movies, the theater presented concerts, plays, and vaudeville acts. The structure still stands today and now serves as an office building. (Courtesy of CRA.)

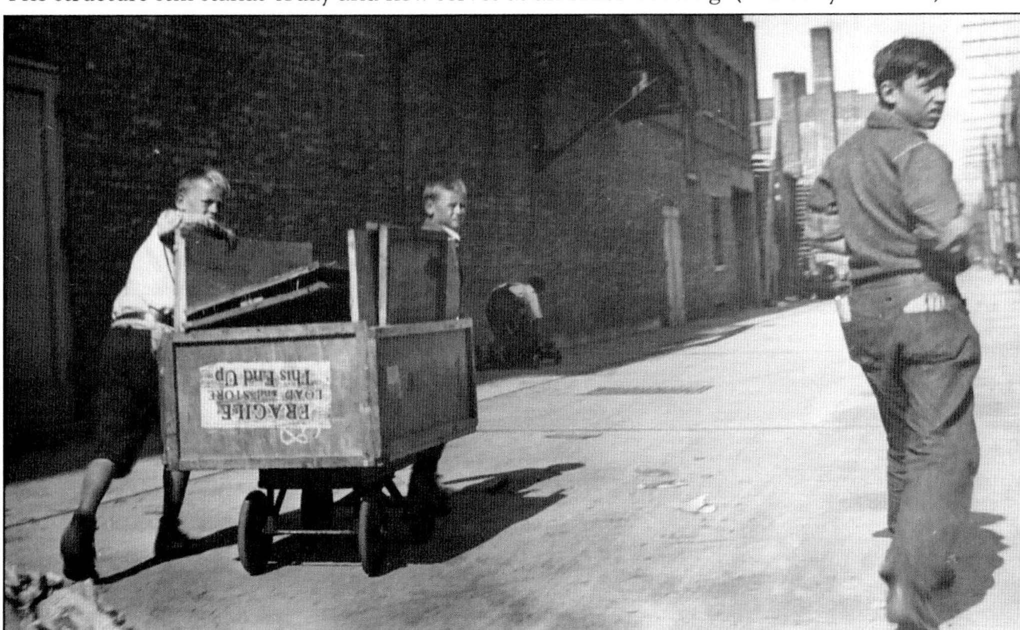

Depression, 1931. During the hard times of the Great Depression, whole households pitched in to help survive. Most homes across the nation were heated by coal, and when coal was too expensive to afford, children were often sent to pick up pieces of coal along the railroad tracks or wood from alleys to keep the furnaces running in winter. Here, youngsters are seen picking up wood in the alley behind the Palace Theater near Eighth Avenue. (Courtesy of CRA.)

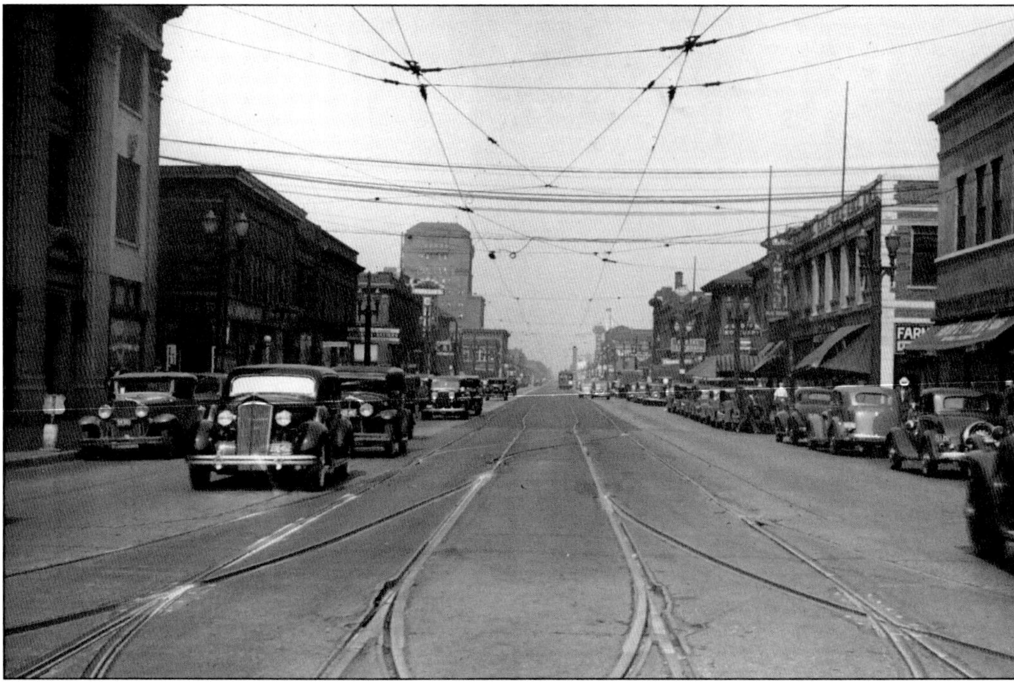

West Fifth Avenue, 1936. Though the United States was still in the midst of the Great Depression, the city of Gary was on its way to recovery. Even with businesses hiring few new employees, few were closing their doors around the time of the 1936 election. This photograph appears to show plenty of business activity along Fifth Avenue west of Broadway. (Courtesy of CRA.)

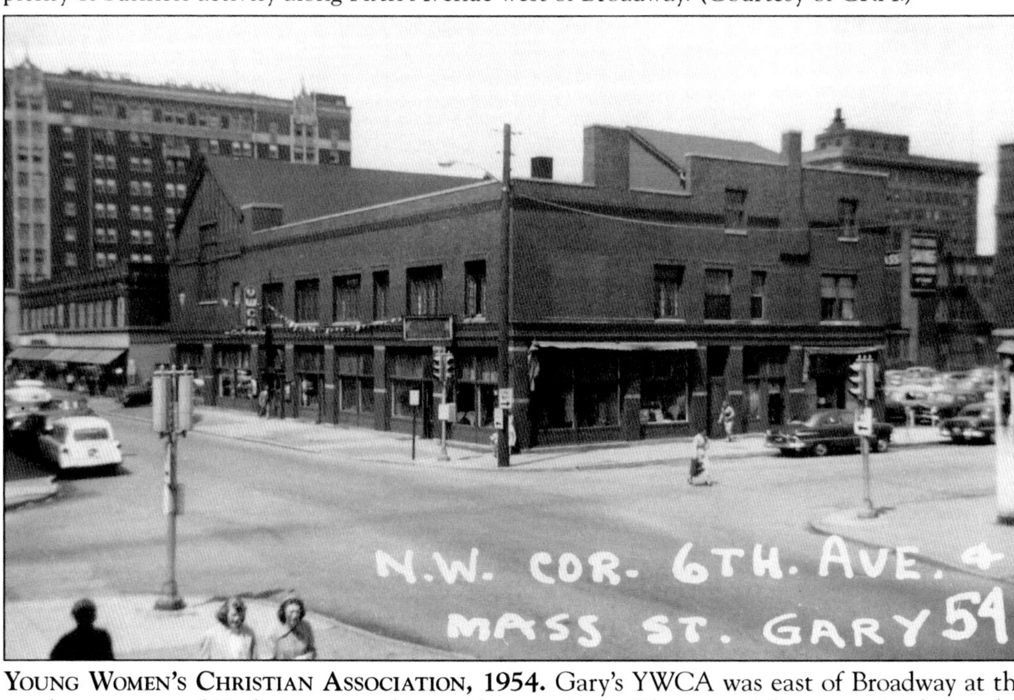

Young Women's Christian Association, 1954. Gary's YWCA was east of Broadway at the northwest corner of Sixth Avenue and Massachusetts Street. The ground floor contained a number of shops, while the upper floor had offices, a dining hall, and a small gym. (Courtesy of CRA.)

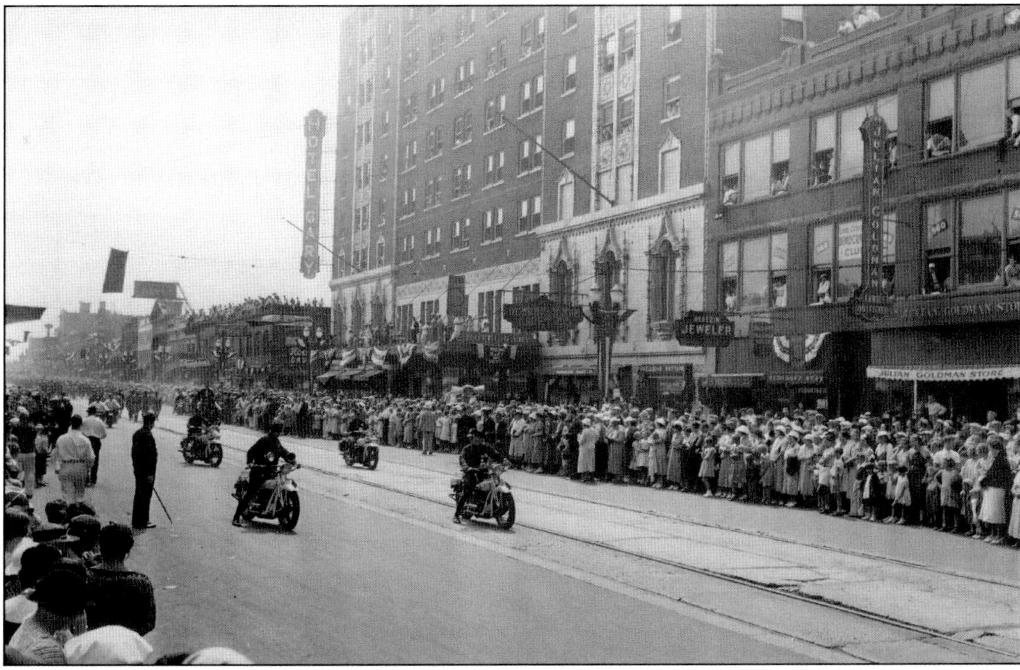

MEMORIAL DAY, 1935. Even though the nation was still in the midst of the Great Depression, Americans did what they could to take their minds off the economic hardships they faced. Some went to free local sporting events, and others, if they had a quarter to spare, went to the movies. On holidays, like Decoration Day (present-day Memorial Day), families attended parades. In 1935, the City of Gary went all out to stage a grand review for the citizens. Pictured are the parade-goers along both sides of Broadway. The photograph above shows the west side in front of the Hotel Gary, while below is the east side. Note the rooftop crowds. (Courtesy of CRA.)

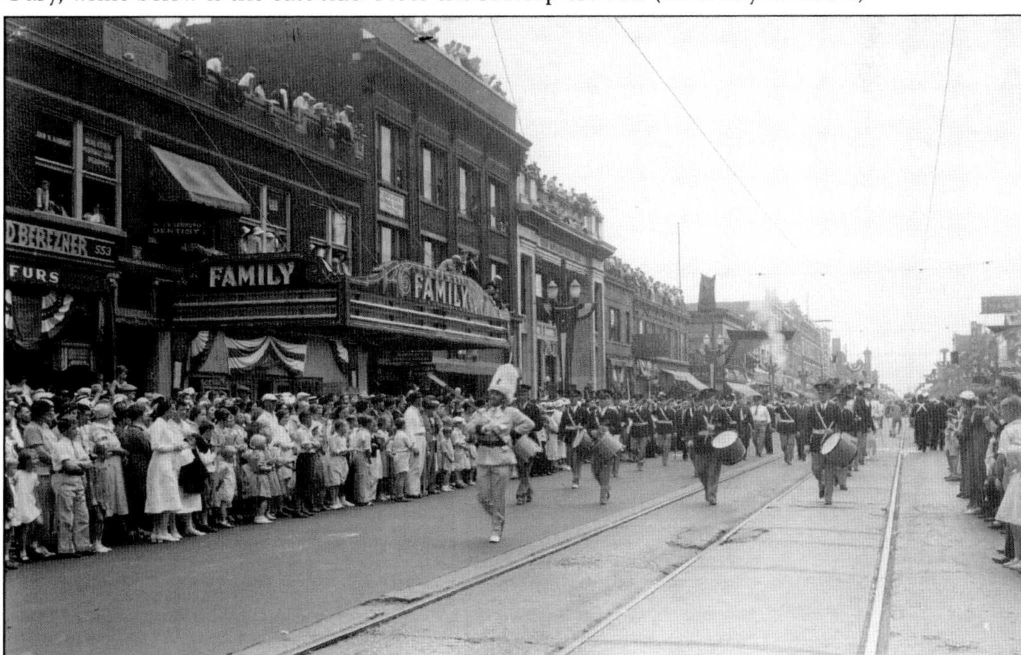

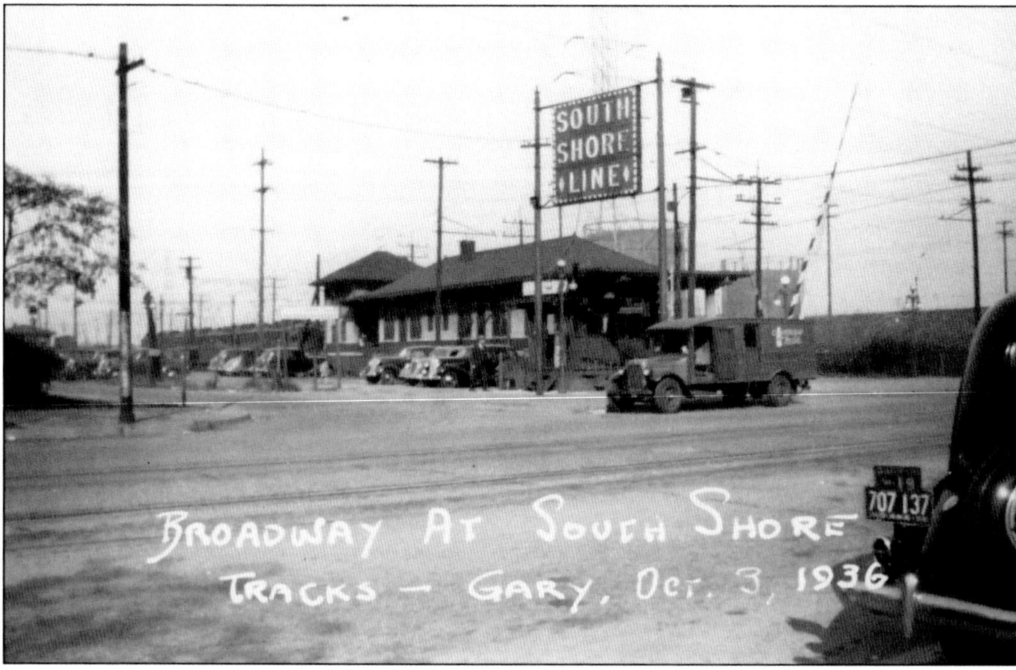

SOUTH SHORE STATION, 1936. The original South Shore Railroad Depot on Broadway is shown here in October 1936. Older riders may remember the orange cars, which were replaced during the 1980s after the railroad survived a number of financial problems that almost took the line out of business. Known as the little train that could, it became the little train that did after it was taken over by the Northern Indiana Commuter Transportation District. The South Shore continues to provide service to Chicago and east to South Bend. (Courtesy of CRA.)

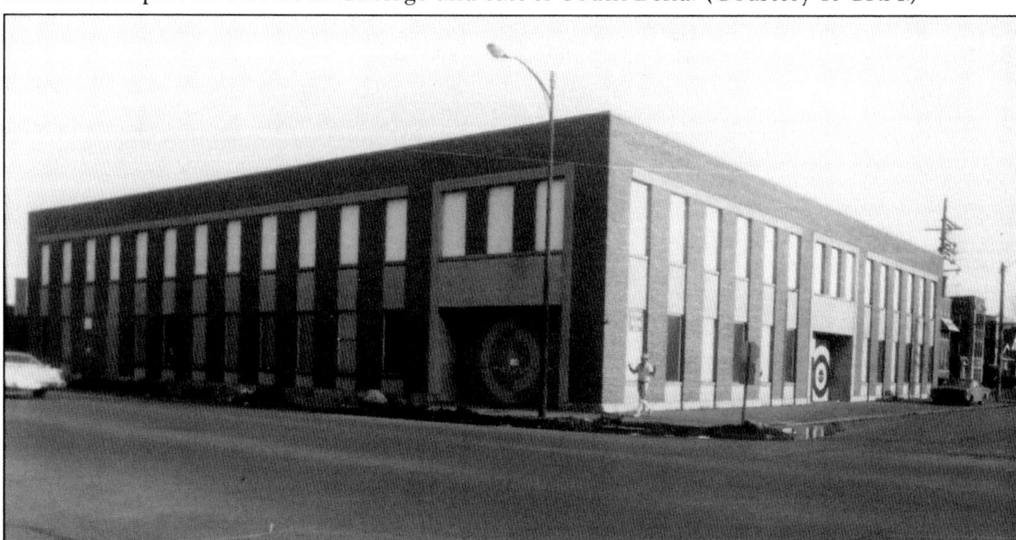

VETERANS' BUILDING, 1970. To meet the needs of the 16.5 million returning servicemen in 1945 at the end of World War II, the federal government constructed new administrative offices for the Veterans Administration across the country. A new facility was built at Fifth Avenue and Pennsylvania Street. It continued operations until 1970, when the government closed a number of federal offices. (Courtesy of CRA.)

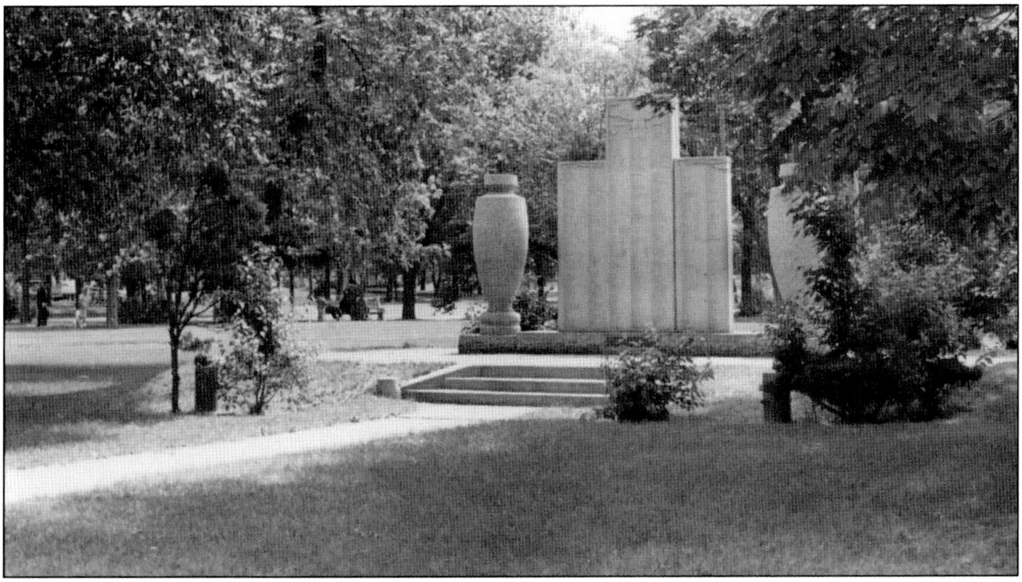

WORLD WAR II MEMORIAL, 1946. A number of monuments and memorials were erected to those who served and sacrificed for the nation in past wars. In Buffington Park near Sixth Avenue is the World War II Memorial, dedicated to the young men of the east side who served in that conflict. The stone monuments list the names of those who served and those who made the ultimate sacrifice. World War II veterans, under the leadership of John Sullivan, restored the neglected memorial in the 1980s. Sullivan was a Navy veteran and retired Gary fireman. (Courtesy of CRA.)

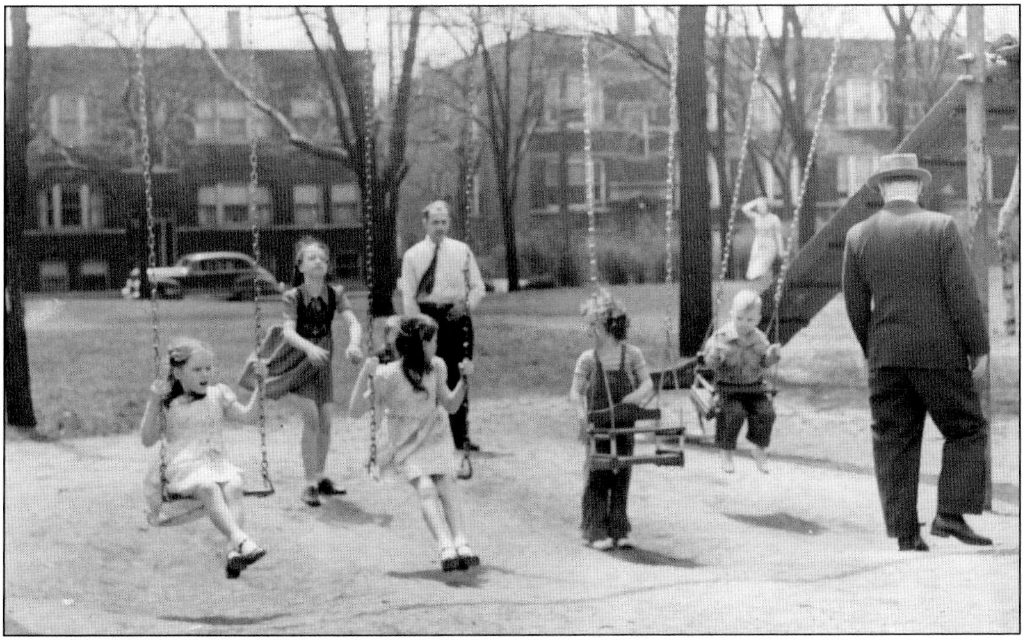

BUFFINGTON PARK, 1947. With the end of World War II in 1945, returning servicemen began new lives as husbands and parents. Here, dads watch their children as they enjoy some playful time on the swings at Buffington Park near Seventh Avenue and Connecticut Street. In those days, swings, slides, and monkey bars were in common use, and if a youngster fell off, he or she got back on again. It was a steel town, and children were tough. (Courtesy of CRA.)

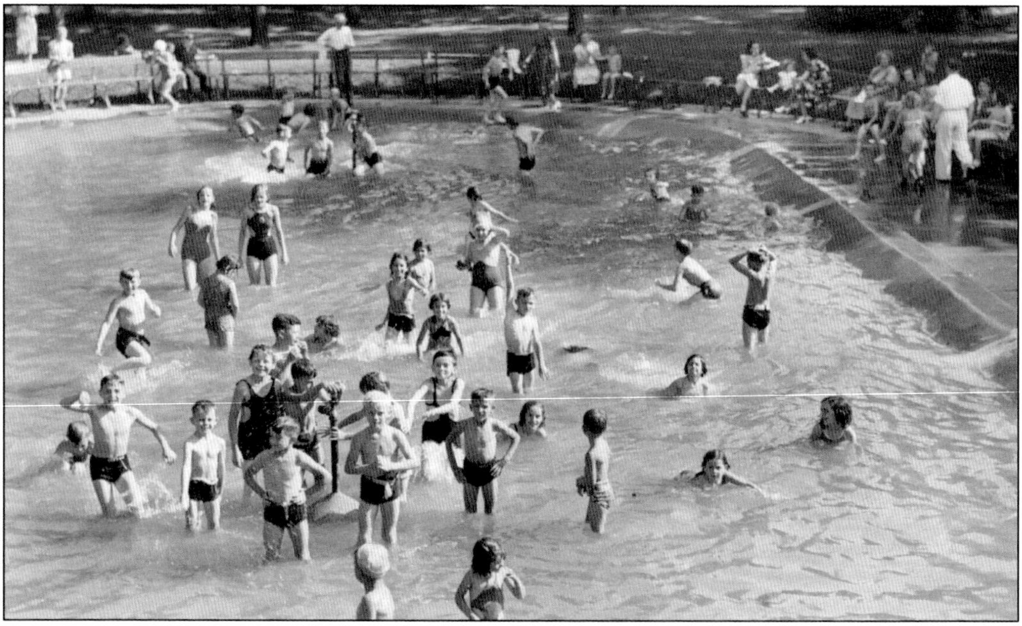

Buffington Park Pool, 1934. In the days before air-conditioning, people used fans to move the air, opened windows, or just suffered through the heat spells. To help Gary residents cool off, US Steel Corporation built neighborhood parks and pools throughout the city. People could also take streetcars to Marquette Park in the Miller neighborhood and seek comfort at the beach along Lake Michigan. The Emerson neighborhood enjoyed access to Buffington Park at Seventh Avenue and Connecticut Street, and the Mann community had Borman Park at Seventh Avenue and Madison Street. Many even slept in the parks at night when the summer heat was unbearable. It was a different and much safer era. (Courtesy of CRA.)

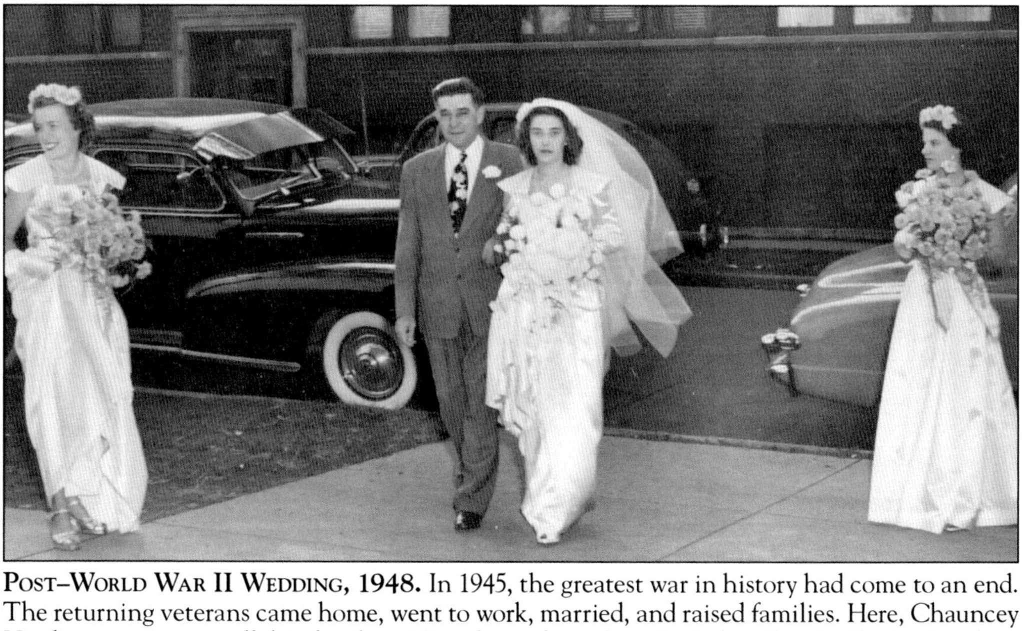

Post–World War II Wedding, 1948. In 1945, the greatest war in history had come to an end. The returning veterans came home, went to work, married, and raised families. Here, Chauncey Hunker prepares to walk his daughter Mitzi down the aisle at St. Luke Church. Her husband-to-be, Tom McMahon, was a decorated World War II veteran. (Courtesy of Colette Oram.)

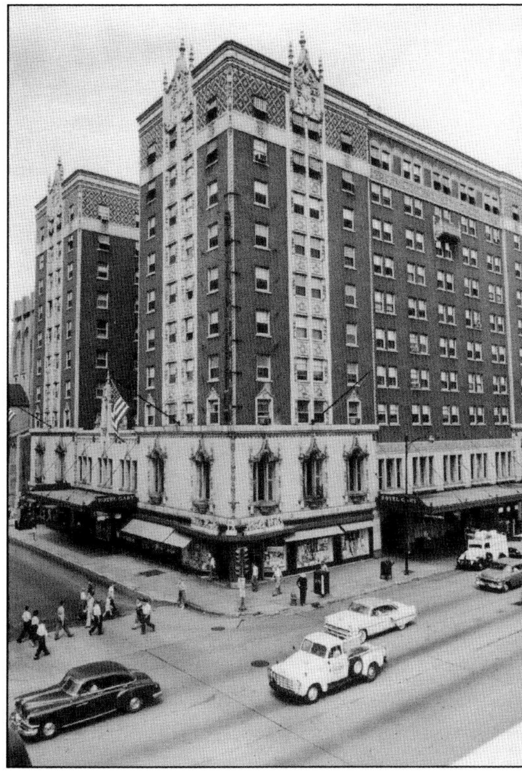

New Hotel Gary, 1950s. Another structure that went up during the 1920s building boom was the Hotel Gary at the northwest corner of Sixth Avenue and Broadway. The historic building contained over 100 guest rooms, shops, restaurants, ballrooms, and meeting rooms. After it closed in the 1970s, it was renovated and reopened as senior housing. (Courtesy of CRA.)

Hotel Gary Dining Room, 1950s. The Crystal Dining Room, located on the second level of the Hotel Gary, played host to many types of social gatherings in its day. Businesses held luncheons and awards ceremonies during daytime gatherings. On many weekends during the school year, local schools held turnabout dances, proms, and sports awards dinners. Pictured is a business gathering during the late 1940s or early 1950s. (Courtesy of CRA.)

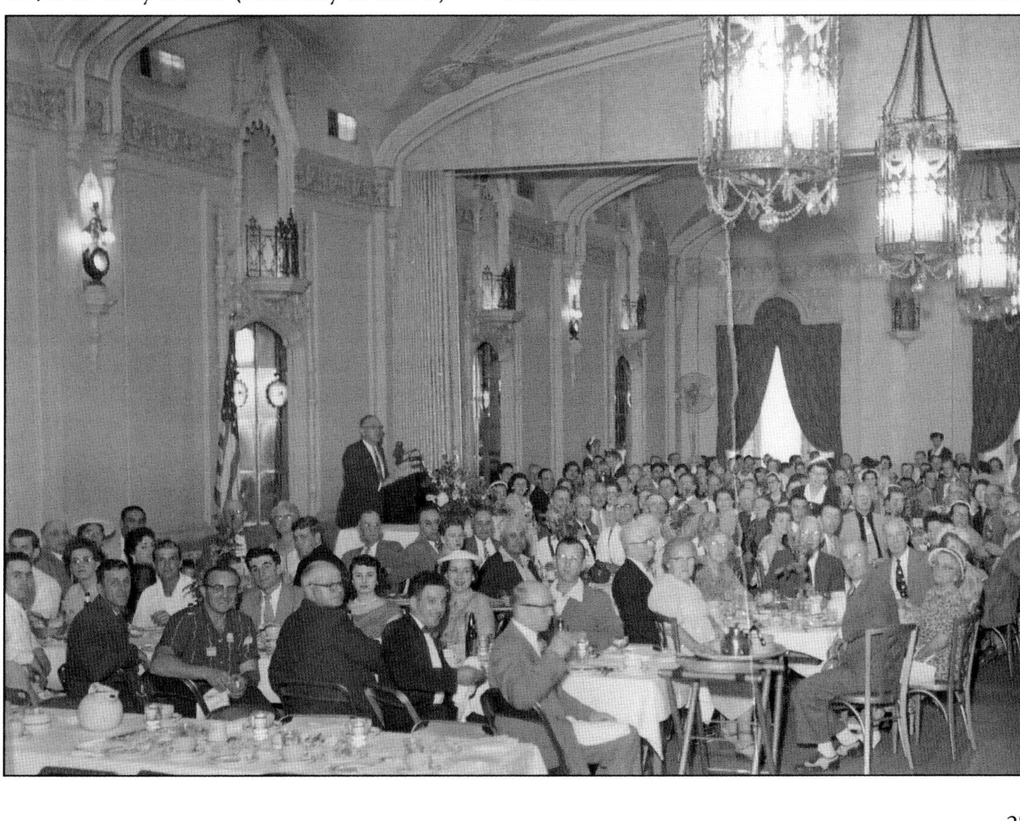

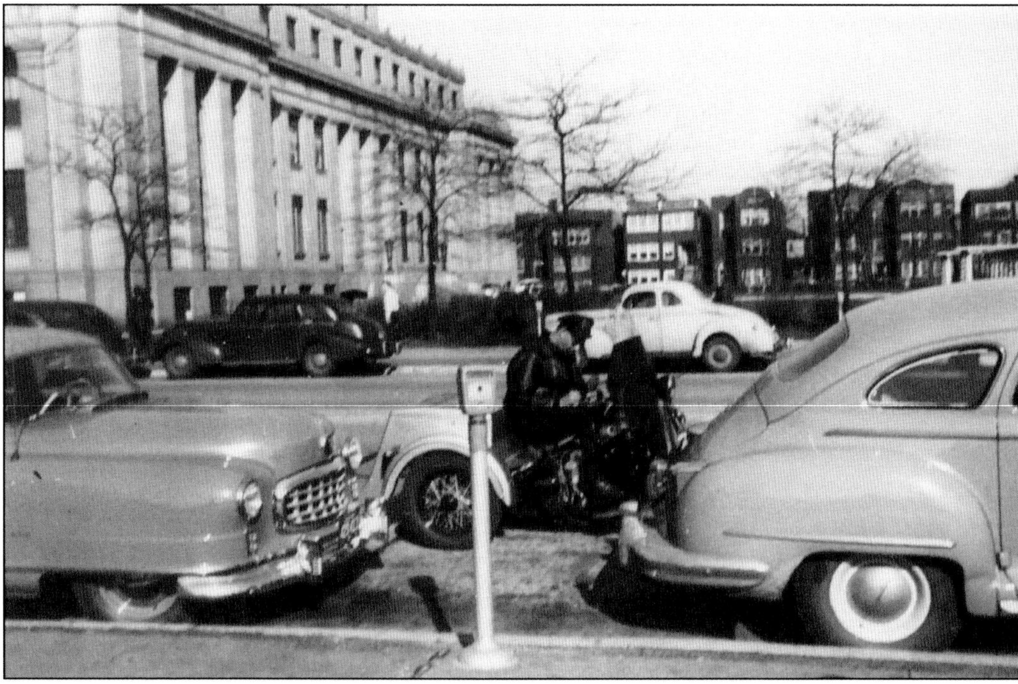

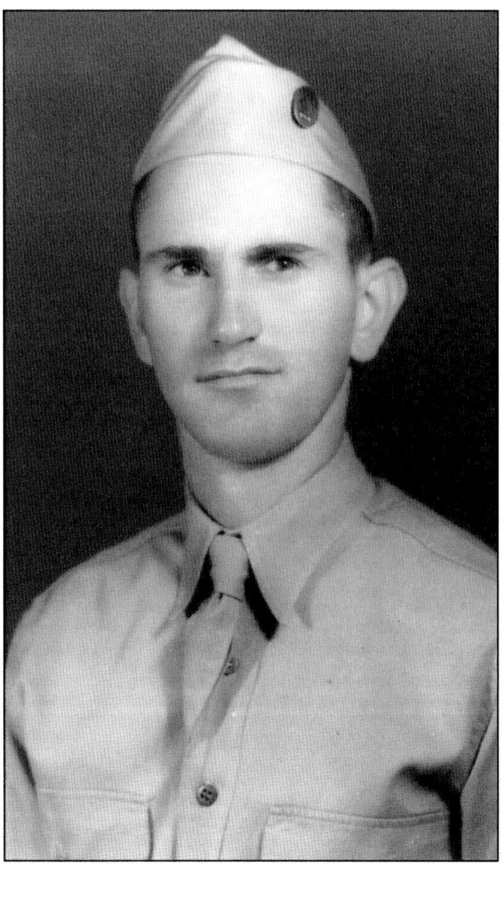

COP GIVING TICKET. With World War II over, the new demands for steel switched from military needs to that of peacetime business. People were better able to purchase automobiles once again. Of course, more cars increased traffic and the need for parking spaces. Pictured is a Gary traffic officer writing a ticket for a car parked past the time allowed by the meter. The auto might have belonged to a shopper or someone conducting business in city hall across the street. (Courtesy of CRA.)

STEVE TRAFNY, 1941. In the fall of 1939, Steve Trafny enlisted in the US Army and was sent to Schofield Barracks in Hawaii. He planned to return stateside in late 1941, but the attack on Pearl Harbor and other military installations changed everything. He served with the 24th Division for the duration of the war and saw action in New Guinea and the Philippines. One of his brothers, John, served in Europe, North Africa, and Italy. Two others, Bill and Stanley, partook in the Normandy invasion and, later, in the brutal Battle of the Bulge. Brother-in-law Charles Burris, a pilot, survived the Bataan Death March and three years as a prisoner of war. This picture was taken in November 1941. (Courtesy of Trafny family.)

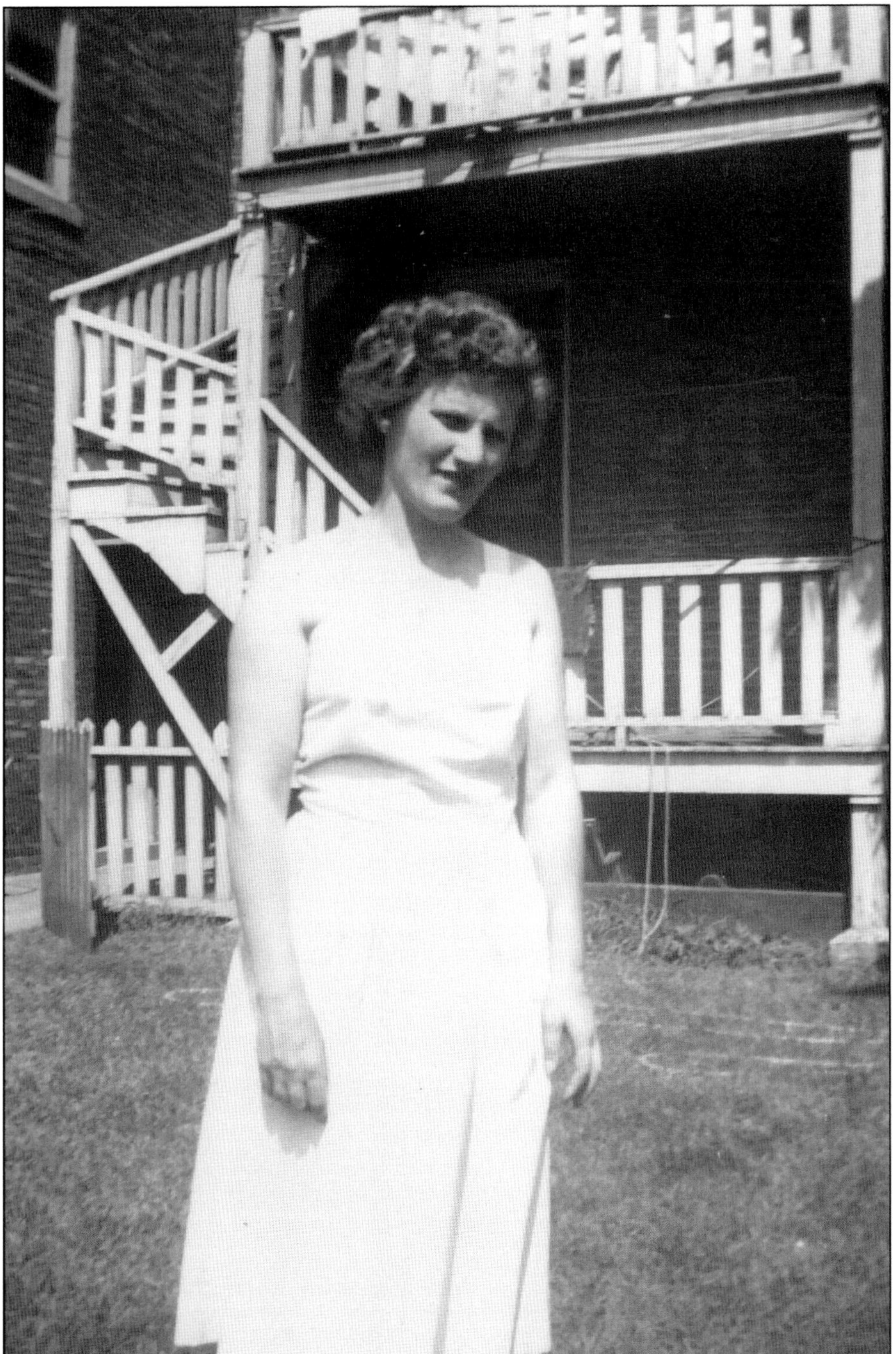

Postwar Bride, 1950. After his service in the South Pacific, Steve Trafny returned home in 1945, but went back into the Army the following year. He was part of the US Army occupation of West Germany. There, he met Wanda Wlodarczyk, a refugee from Poland. She was sent to forced labor in Germany and witnessed the brutality of the German war machine in Poland firsthand. Steve and Wanda were married in 1948, and settled on Gary's east side. Wanda is pictured in 1950. (Courtesy of Trafny family.)

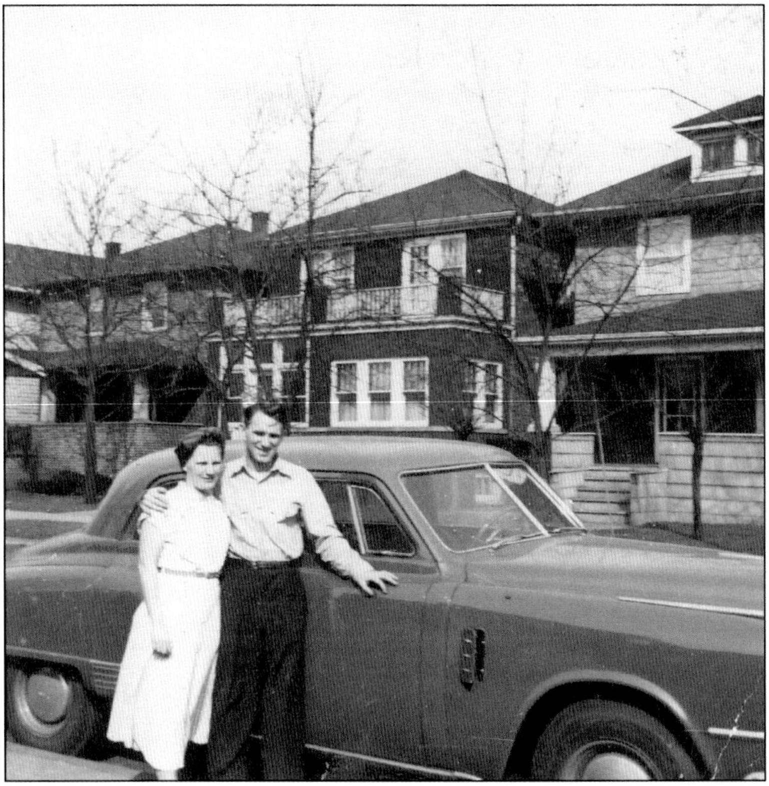

Steve and Wanda, 1950. Steve and Wanda Trafny are pictured in 1950 at their home at 832 Carolina Street. They later moved to 780 Indiana Street. The car belonged to his sister and brother-in-law, Helen and Philip Lis. (Courtesy of Trafny family.)

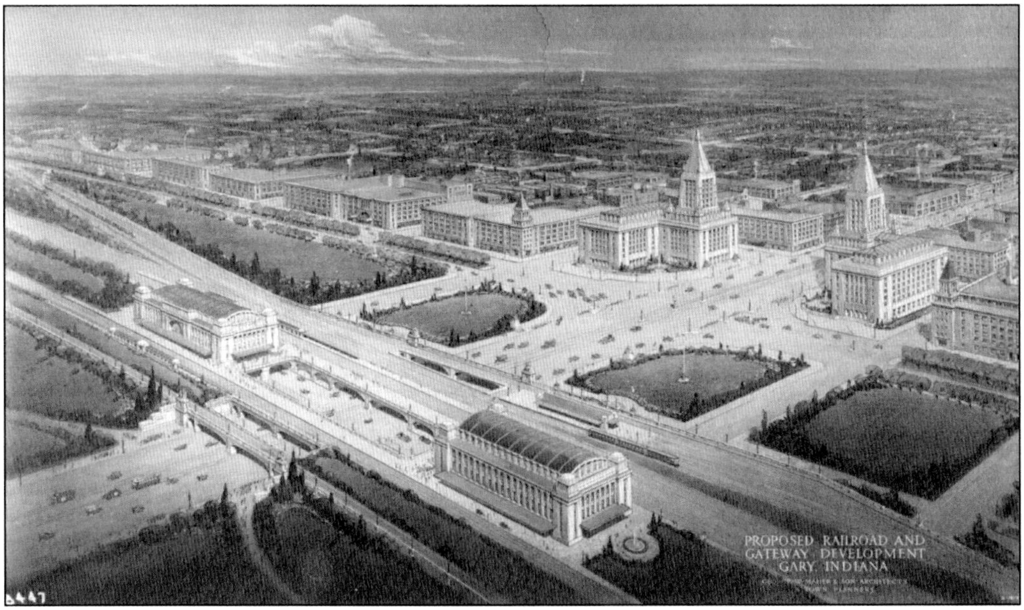

Gary Development Plan, 1924. Like most major American cities, Gary did have plans for future development and expansion. In 1924, a railroad and gateway development plan was proposed by George W. Maher & Son, architects and town planners. The design shows Downtown Gary at Fourth Avenue and Broadway. This proposal became outdated with the arrival of the Interstate Highway System and suburbia in post–World War II America. (Courtesy of CRA.)

Two

US Steel Corporation and Unions

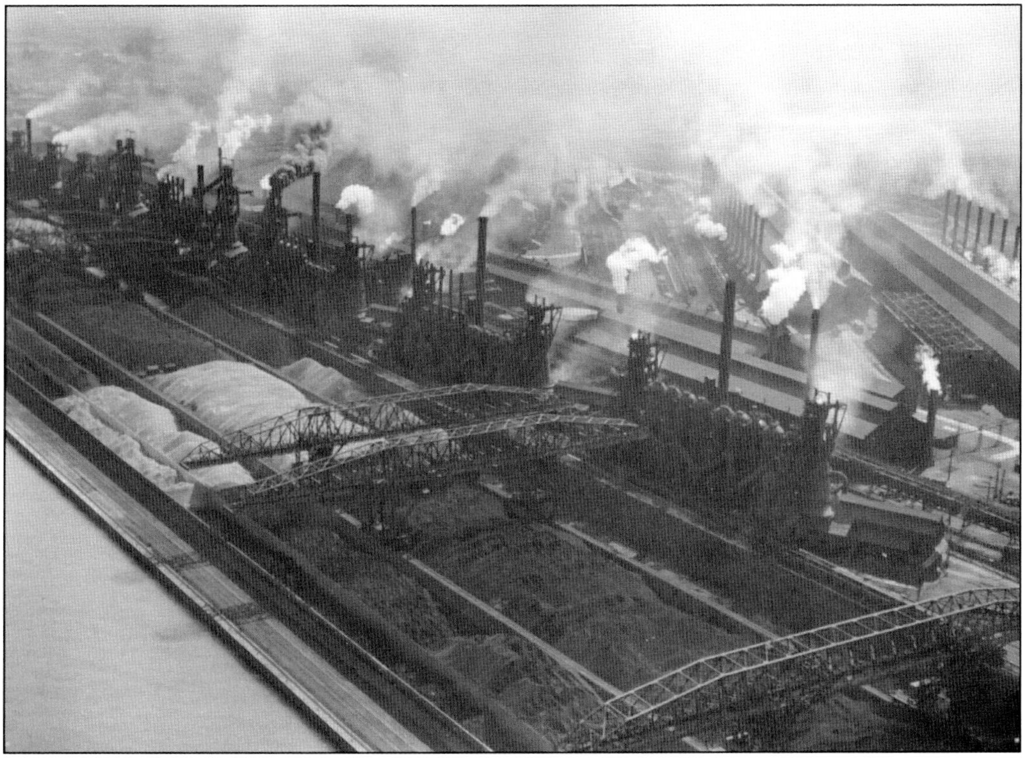

Aerial View of US Steel. In the early years before the unions, men worked in grueling and often unsafe conditions. They worked 12 hours a day, six days a week, in a hot, dirty environment. Every two weeks or so, they all worked a 24-hour swing shift for no overtime pay. Early attempts at unionization were sometimes met with violence, loss of jobs, and blacklisting. By the late 1930s, the company recognized the Steel Workers Organizing Committee union. The men and women who worked at the Gary Works made the steel that was used to win two world wars, help rebuild Europe, and make the steel beams for the highway bridges that would be part of the Interstate Highway System that began under President Eisenhower in the 1950s. In addition, the Gary plant made the steel for the beams that were used to make the skyscrapers that are now part of Chicago's skyline. Even the Chicago Picasso sculpture had its roots in the Gary plant. Though wages and conditions have improved over the years, the steel-making process is still dangerous. For the millrats in the blast furnace area, coke plant, basic oxygen steelmaking (BOP) shop, and other departments at the Gary plant, working there is always a challenge. (Courtesy of CRA.)

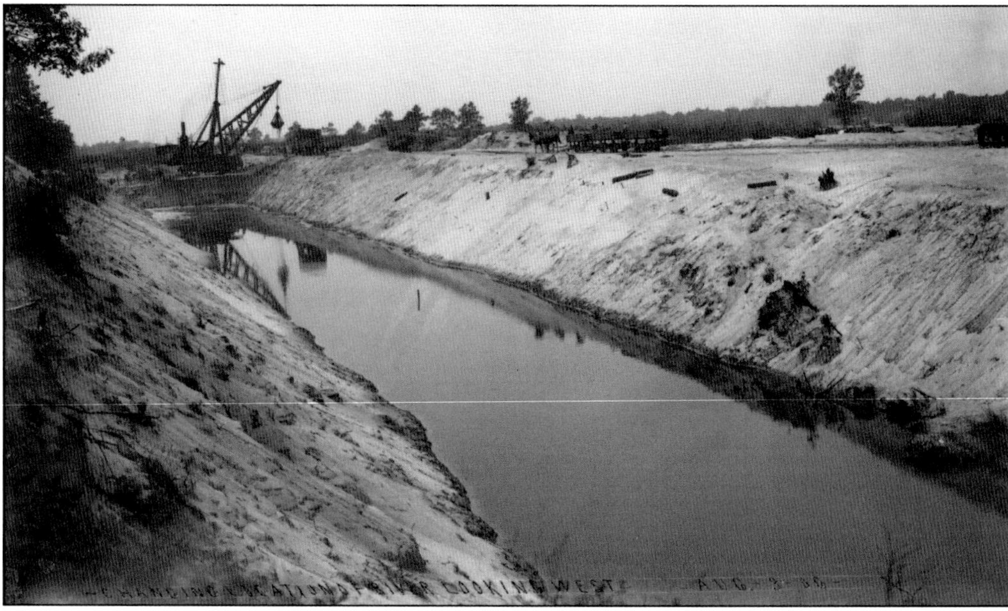

Moving the Grand Calumet River, 1906. One of the first major construction projects that US Steel undertook in 1906 was to relocate the Grand Calumet River some 300 yards south of its original location, which cut through the mill property. Once the river project was completed, the new channel formed a dividing line between the town and plant property. Pictured is a giant steam-powered crane moving sand along the riverbed. (Courtesy of CRA.)

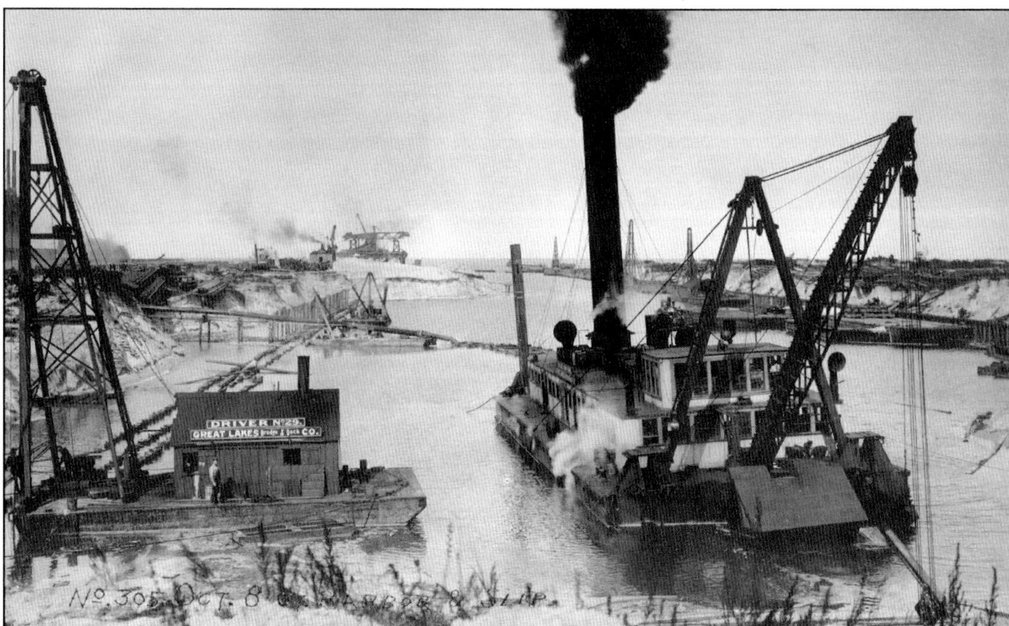

Digging the Channel, 1906. In addition to moving the Grand Calumet River, the steel company had to dig a channel for the ore boats to dock and unload their cargoes of iron ore and limestone. Most of the coal needed in the steel-making process came by rail. Pictured are the digging operations in 1906. In 1908, the first ore boat, the *Elbert H. Gary*, unloaded the first cargo for the blast furnaces. (Courtesy of CRA.)

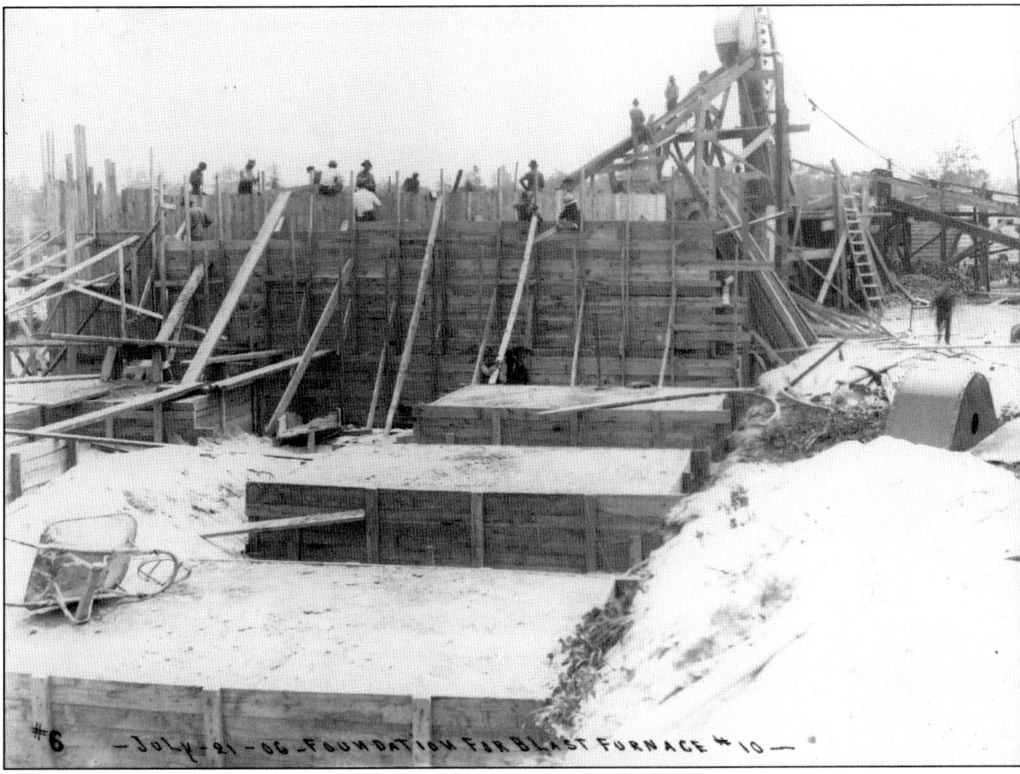

BLAST FURNACE FOUNDATION WORK, 1907. Beginning in 1906, ten modern blast furnaces were constructed at the plant site, and within two years, iron-producing operations commenced. In December 1908, No. 8 Blast Furnace poured the first iron for steel production. Members of the construction crew are pictured laying the foundation for a new furnace. (Courtesy of CRA.)

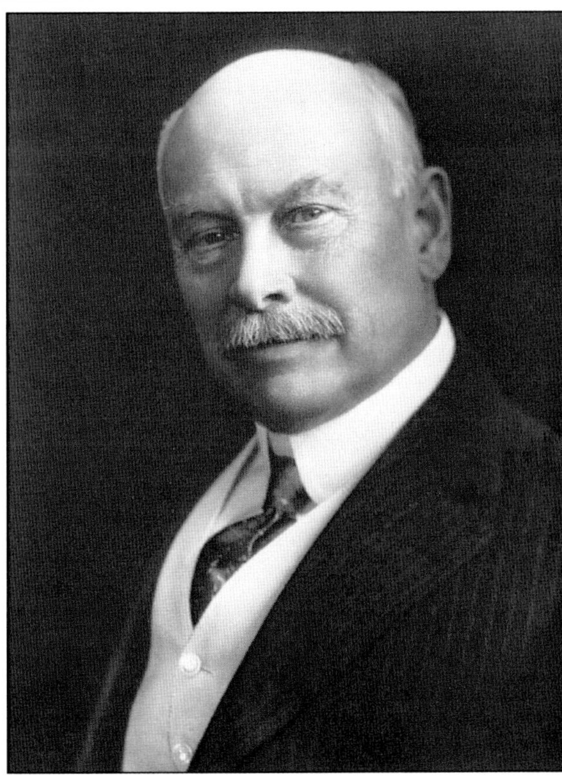

ELBERT H. GARY, 1906. Elbert H. Gary, former judge of DuPage County, became head of Federated Steel in 1898. Three years later, he merged his operations with other companies to help form US Steel Corporation, which became America's first billion-dollar corporation. At first, Judge Gary was not in favor of building the proposed plant in Indiana and instead wanted to locate it in Illinois. However, lower taxes and a location near major railroad and shipping lines convinced him to locate the mill in Indiana. (Courtesy of CRA.)

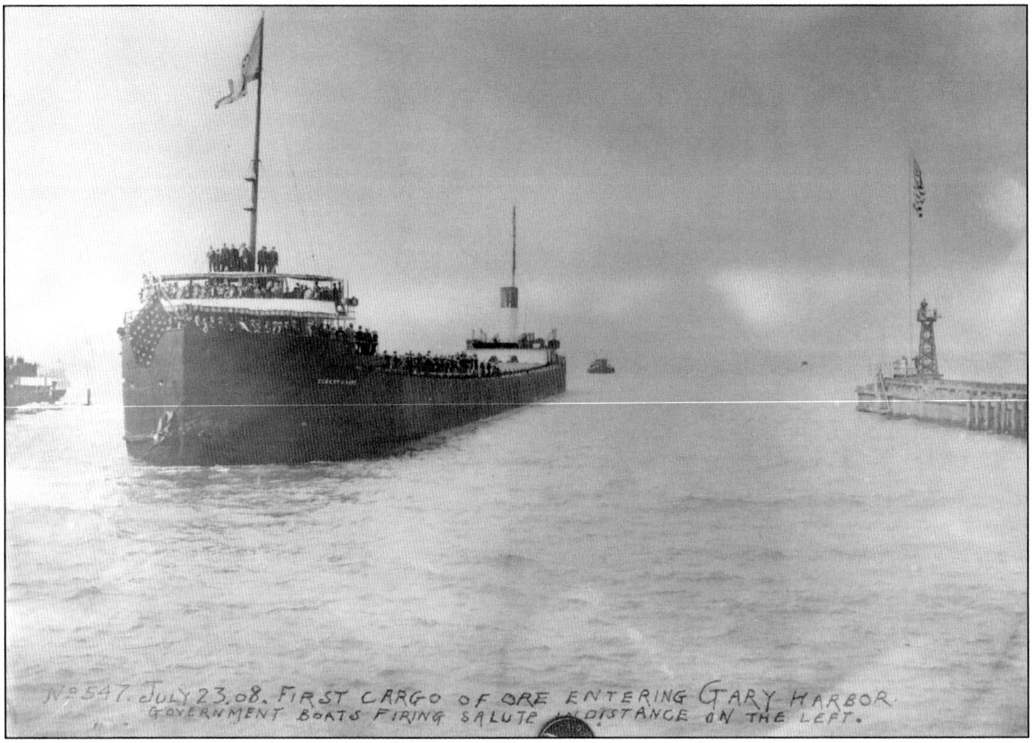

First Ore Boat, 1908. On July 23, 1908, two years after construction began on the Gary plant, the first ore boat, the *Elbert H. Gary*, entered US Steel Harbor carrying the first shipment of iron ore. Coast Guard ships and many private boats were on hand to offer salutes. Hundreds of workers were also present along the docks to take part in the festivities. (Courtesy of CRA.)

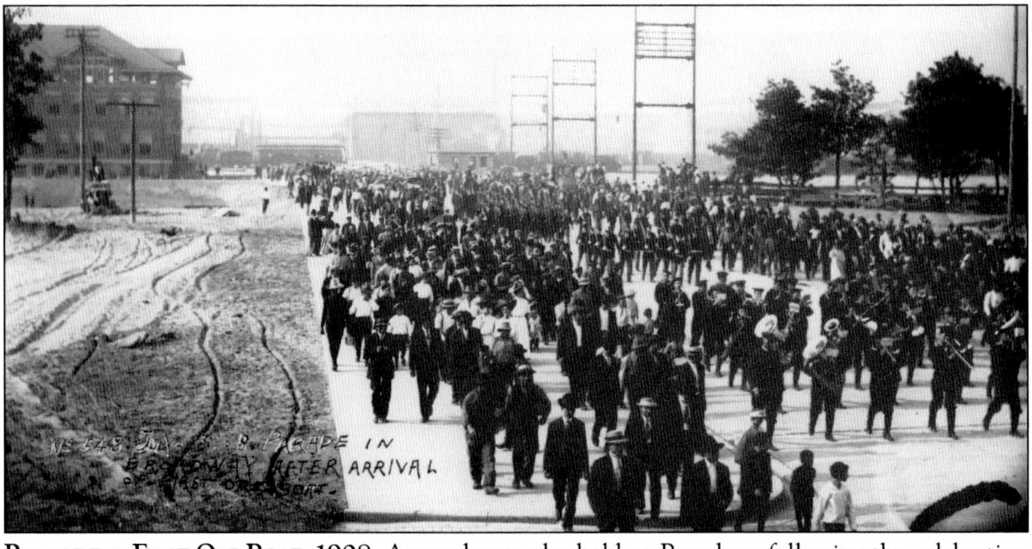

Parade for First Ore Boat, 1908. A parade was also held on Broadway following the celebration on the docks. Various bands marched from the main gate of Gary Works south to Fifth Avenue and Broadway downtown. Of course, workers were required to take part in the celebration if they were not working their shift that day. Company men checked attendance, as the men had no union. (Courtesy of CRA.)

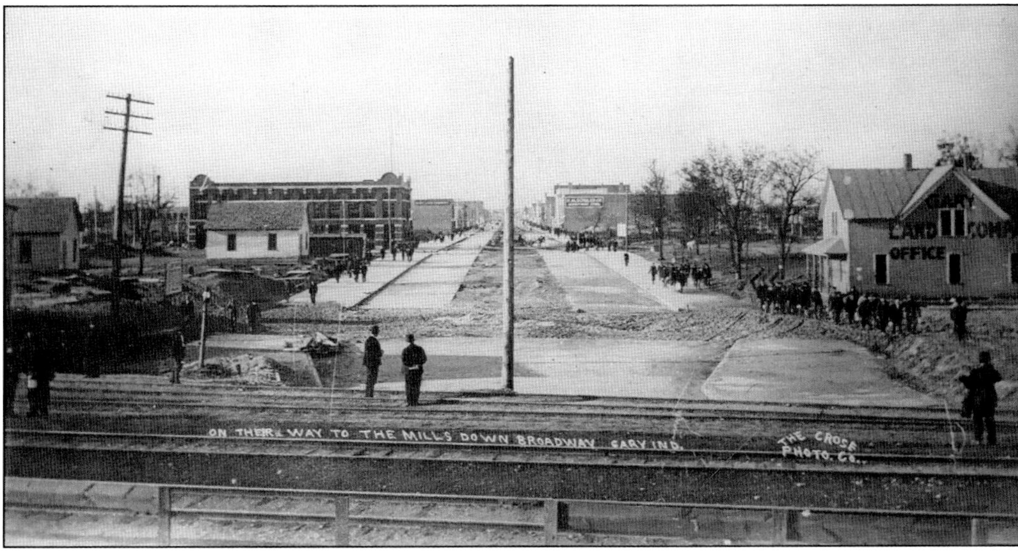

Steelworkers Going to the Mill, 1908. In 1906, US Steel began construction of a modern plant along the shore of Lake Michigan that gave birth to the city of Gary, Indiana. Within a year, the mill was operational and producing steel. In this 1908 photograph, workers are on their way to the day shift at the Broadway gate. In those days, workers had no union to back their interests, so 12-hour shifts, six or seven days a week, were common. Every two weeks, they were forced to endure a 24-hour swing turn as their schedule switched from day to night. (Courtesy of CRA.)

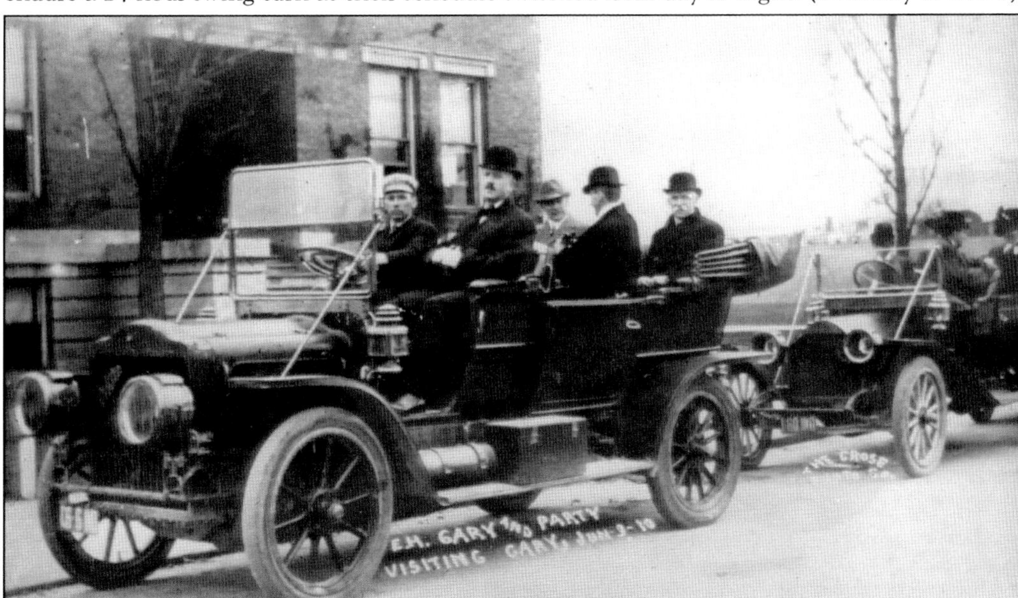

Judge Gary's Visit, 1909. Judge Gary made few visits to the town that bore his name. In 1909, he paid the young and growing steel town a visit. Though the judge lived in New York City, he claimed that his heart was in Gary. He made the Gary Land Company set aside space for parks and churches and allowed his workers, nearly a quarter of a million, to form unions. After losing a bitter strike in 1919, many steelworkers compared him to Kaiser Wilhelm II of Germany. He is pictured here during his 1909 visit. (Courtesy of CRA.)

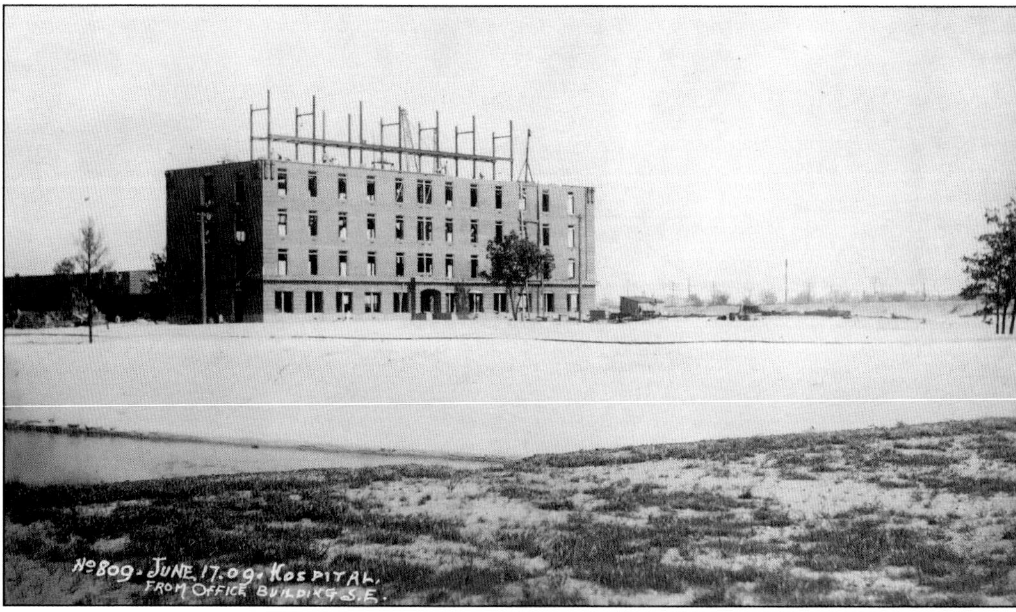

HOSPITAL CONSTRUCTION, 1908. Those who worked in the steel industry during the early 20th century understood the risks of the job in no time. Men worked long hours in dangerous environments where one could face permanent injury or death. US Steel constructed its own hospital on Third Avenue on the west side of Broadway. The hospital is pictured here under construction in 1908. (Courtesy of CRA.)

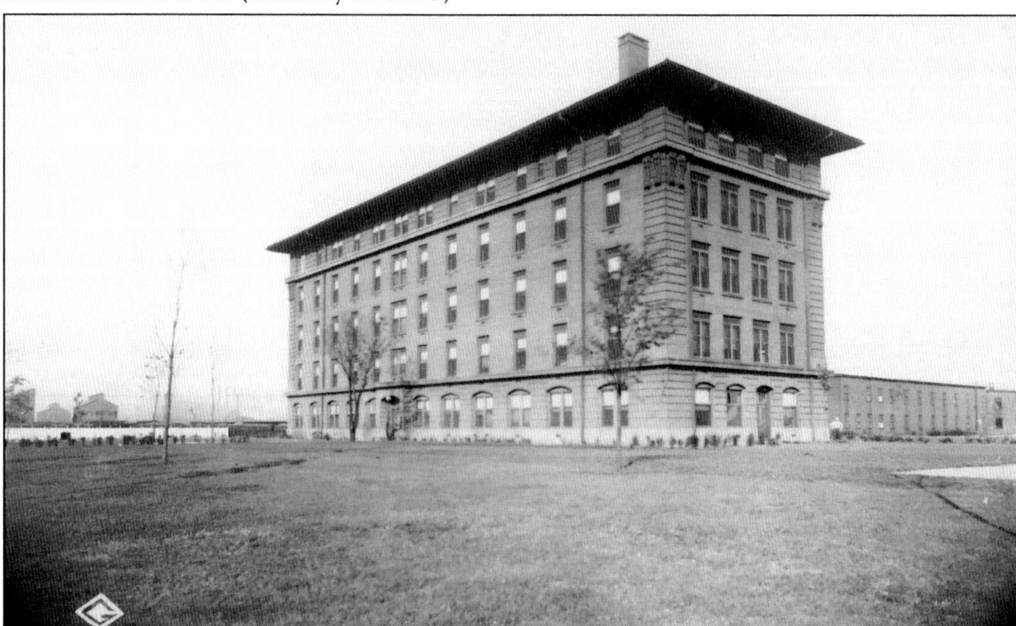

HOSPITAL, 1912. During the early years of operation, the doctors and nurses of the Gary Works hospital cared for sick and injured workers. The Mercy and Methodist hospitals were blocks away, while the mill hospital was just outside the main gate at Broadway. By the late 1920s, injured workers were taken by ambulance to the city hospitals. The mill hospital is still in use today but does not handle serious cases nor house patients. (Courtesy of CRA.)

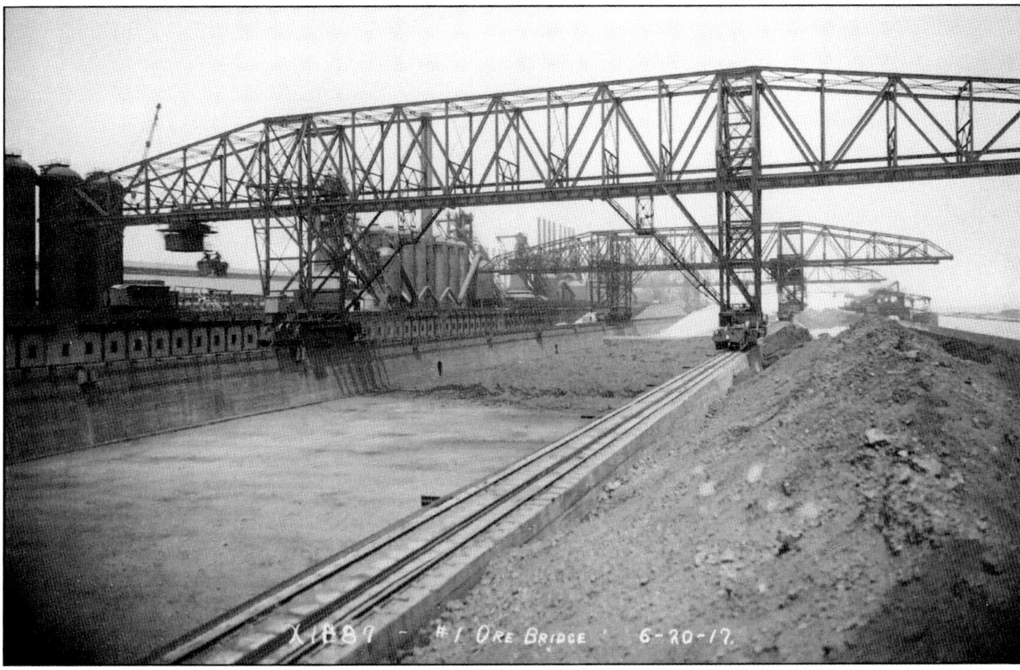

Ore Bridge, 1917. Once the ore boats brought in the iron ore and limestone, specially built ship unloaders moved the cargo to bridges, like the one pictured, and moved the material toward the blast furnace storage bins. The bridges moved along special rails north and south. The bridges were each about a city block long and had to be secured before strong winds and storms. (Courtesy of CRA.)

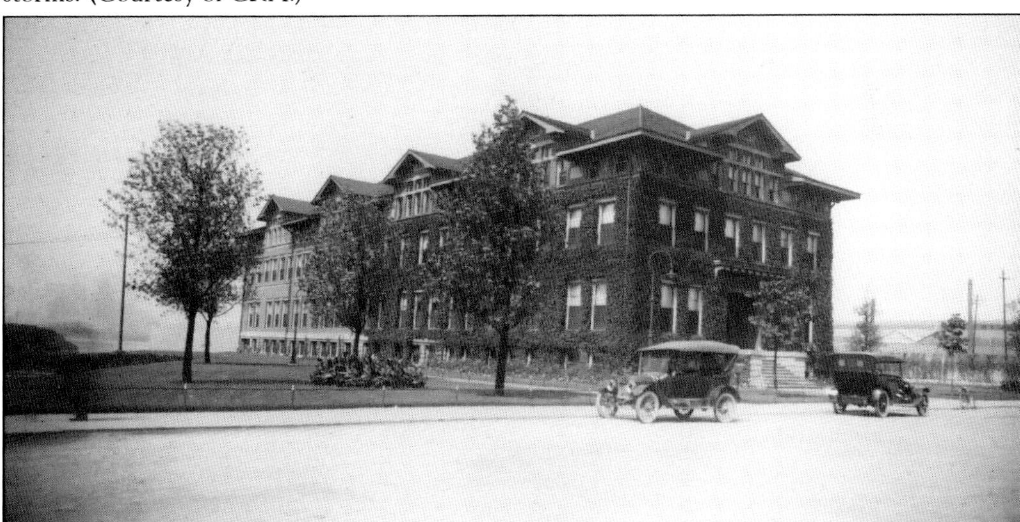

US Steel Main Office, 1917. Once ground was broken for the Gary plant, one of the earliest structures completed and operational was the main office, located on the west side of Second Avenue and Broadway. There were no banks to hold cash in the city of Gary in 1906, so special vaults were shipped and installed on the lower level when the foundation was laid. The employees were paid in cash, so the money had to be held in secure areas. The building is shown here during the winter of 1917. (Courtesy of CRA.)

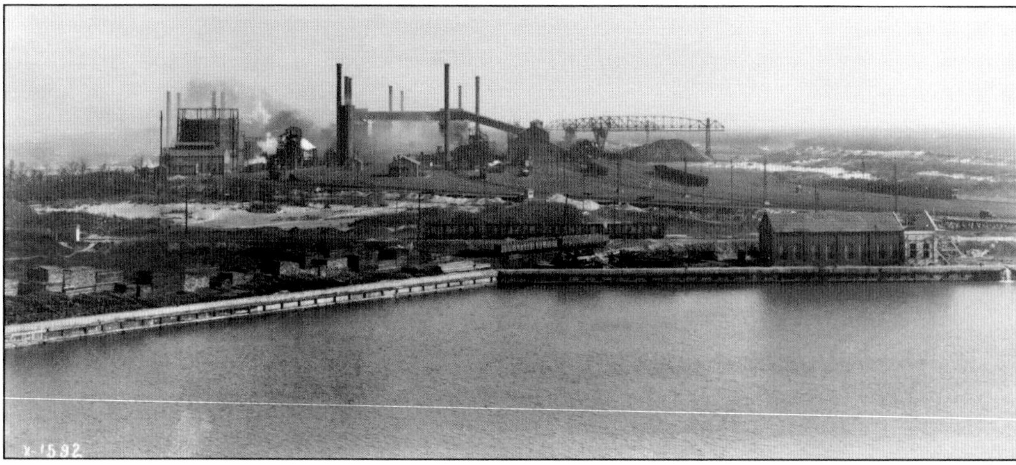

Coke Plant, 1917. The coke plant was built at the eastern end of the Gary Steel Works. There, coal was converted to coke, a carbon product used in blast furnaces to remove the impurities from iron and steel. During the coke process, heat and steam are used to power the blast furnaces. The plant is pictured here during 1917. (Courtesy of CRA.)

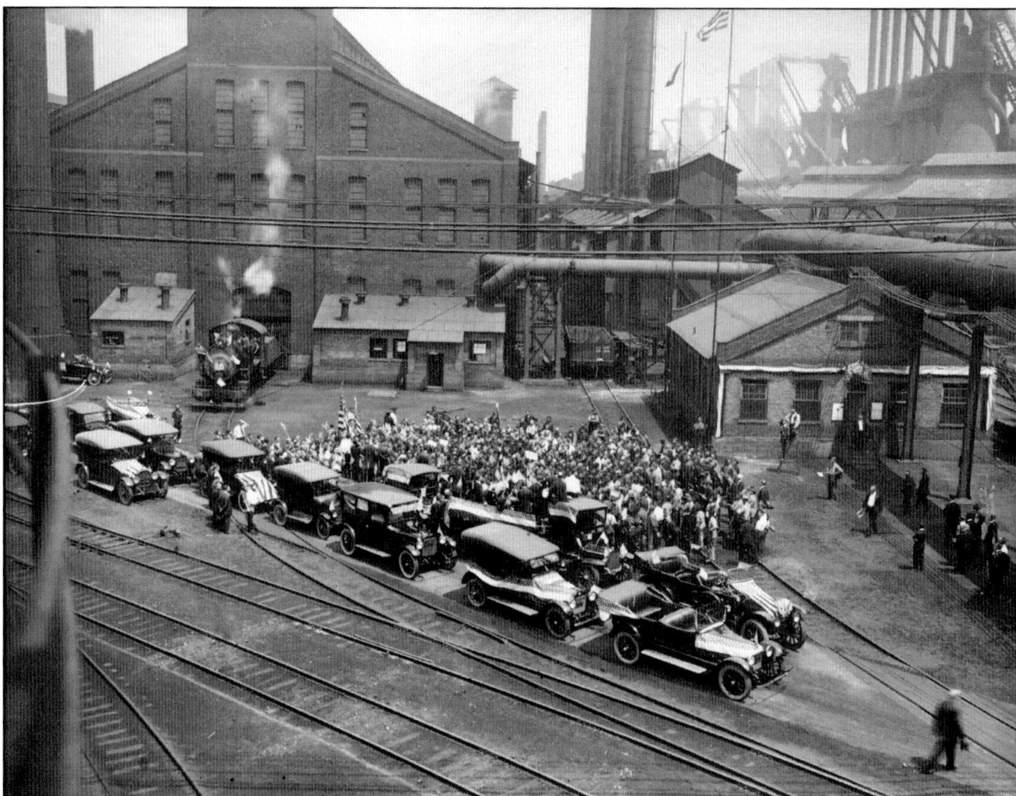

World War I Ceremony, 1918. During America's war years of 1917–1918, countless patriotic rallies, parades, and bond drives were held across the country. At US Steel, such events were common to encourage the workers to do their part to win the war and beat the Kaiser. On April 14, 1918, Flag Day, a ceremony honoring those who were fighting on the Western Front in France took place in front of the chemical lab near the blast furnace department. (Courtesy of CRA.)

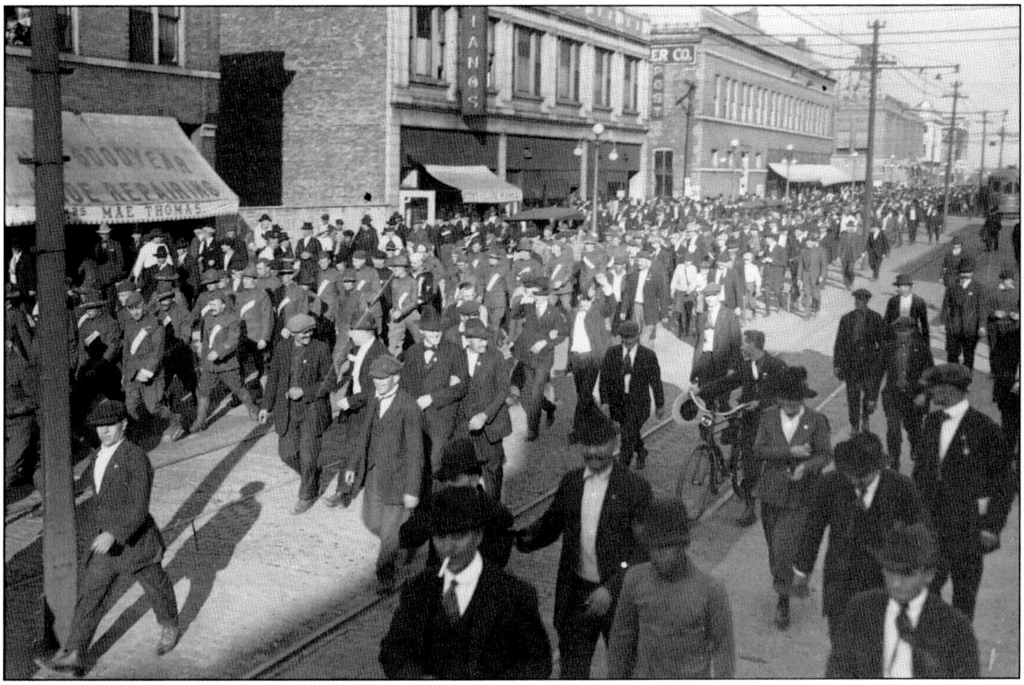

1919 Steel Strike. In 1919, the steel and coal industries were hit with nationwide strikes, and on the West Coast, the longshoremen walked off and shut down the ports. In Gary, the steelworkers' union called for an eight-hour day, six day week, and collective bargaining rights. The company denied their demands, so the workers walked out on strike in early October 1919. When Mayor William Hodges's ban on public protests was ignored by the workers and citizens, he called for the Indiana militia to restore order. Workers still marched on Fifth Avenue. They are pictured here, with some wearing their Army uniforms from the Great War. (Courtesy of CRA.)

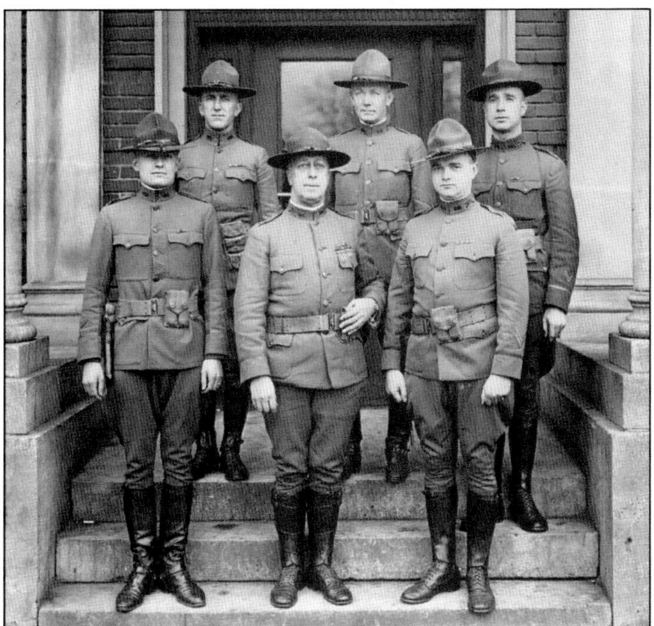

Colonel Mapes and Staff, 1919. Pictured are members of the staff of Col. William Mapes at the main library on West Fifth Avenue during the strike. The troops were requested to keep order in Gary, but on a number of occasions, troops violated the rights of strikers and caused property damages. (Courtesy of CRA.)

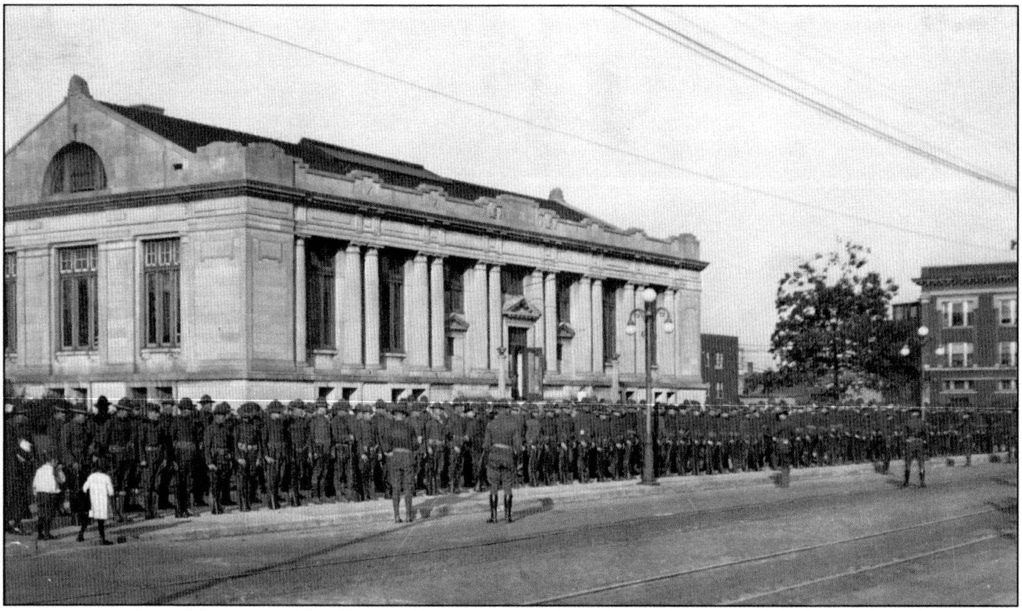

Troops March into Gary, 1919. When strikers ignored Mayor Hodges's order banning any public gatherings, the city was placed under martial law. Gov. James Goodrich requested federal troops to keep order. Gen. Leonard Wood dispatched 1,500 troops from the US Army's 4th Division under Colonel Mapes. Troops are pictured here marching on Broadway near Sixth Avenue. (Courtesy of CRA.)

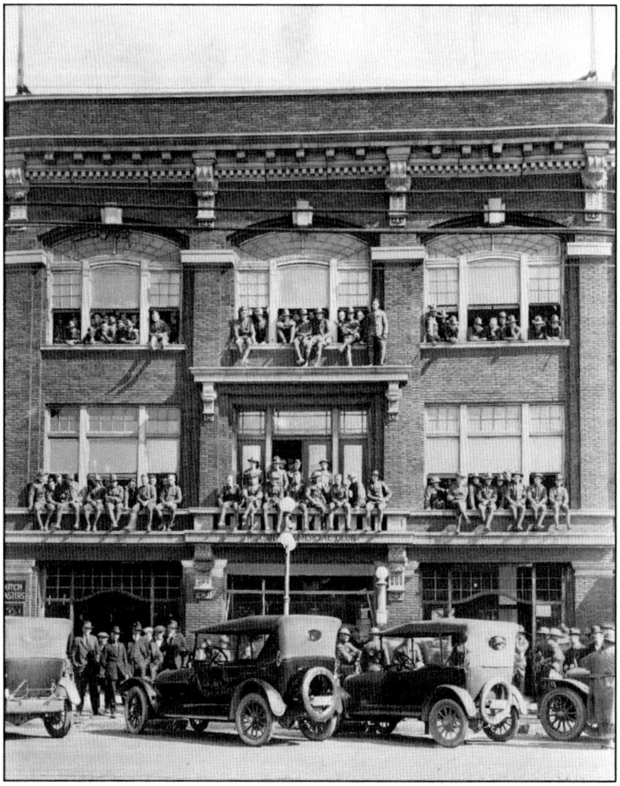

Troops of Commercial Club, 1919. Troops of the 4th Division were housed in the downtown area during the steel strike of 1919. They are pictured here at the Gary Commercial Club in October. The troops were given arrest powers but abused them as they took strike leaders into custody without warrants. They also were guilty of extensive property damage to the building and were never charged by the owners. (Courtesy of CRA.)

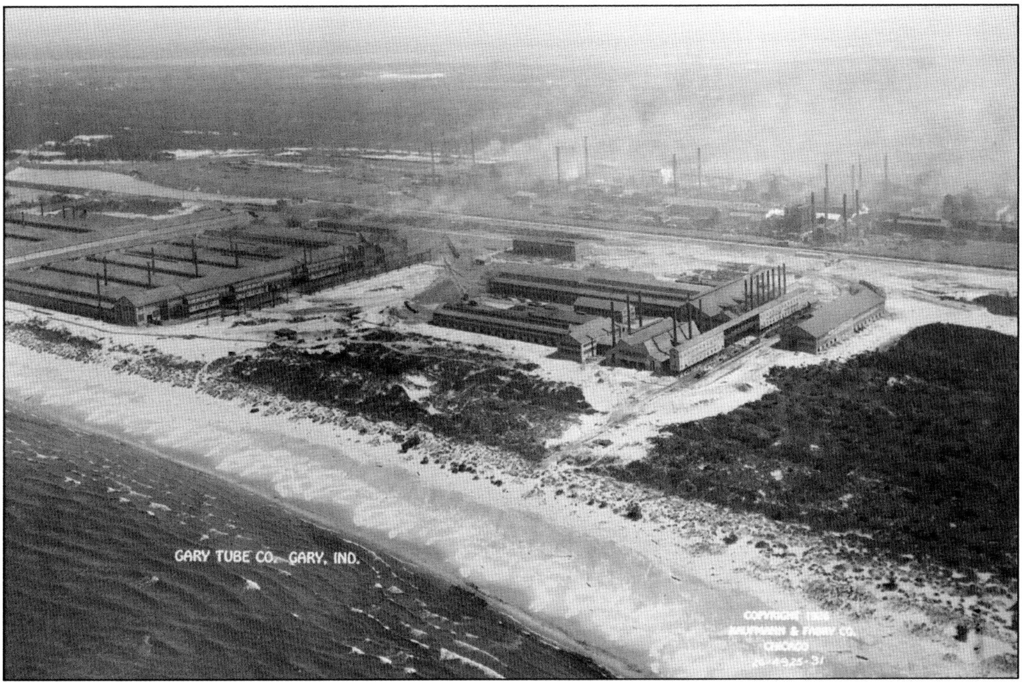

National Tube Works, 1920s. The National Tube Works, a subsidiary of US Steel Corporation, was located at the eastern end of the massive Gary plant. At the time, there was a great demand for steel tubing products, and Gary Works had the resources to provide the finished goods on time and at competitive prices. The plant is pictured in 1926, just north of the coke plant. (Courtesy of CRA.)

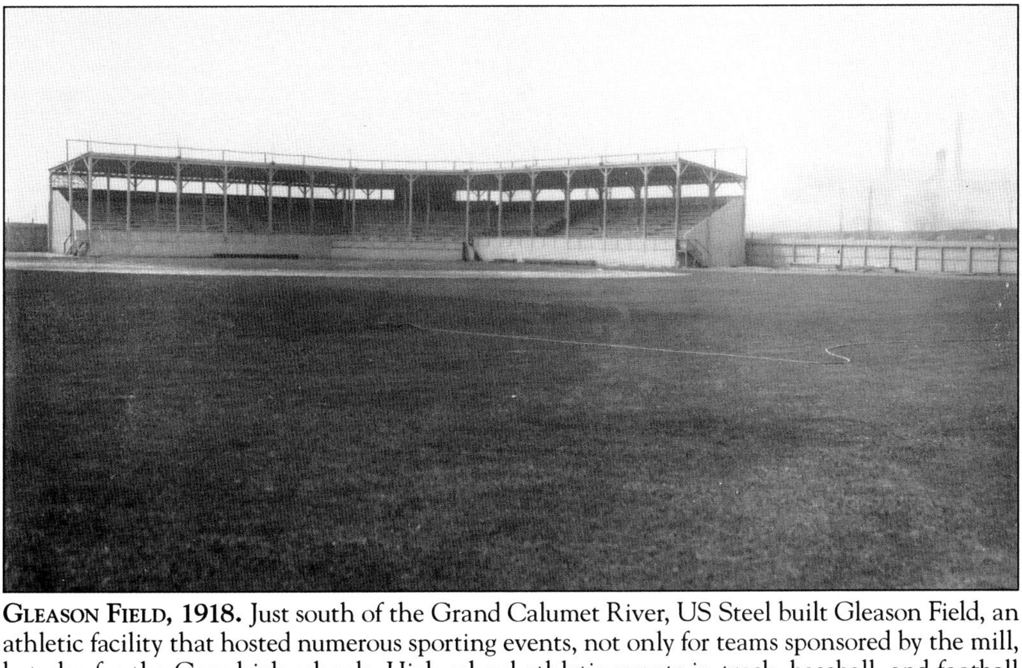

Gleason Field, 1918. Just south of the Grand Calumet River, US Steel built Gleason Field, an athletic facility that hosted numerous sporting events, not only for teams sponsored by the mill, but also for the Gary high schools. High school athletic events in track, baseball, and football took place at the facility. It is pictured here around 1918. (Courtesy of CRA.)

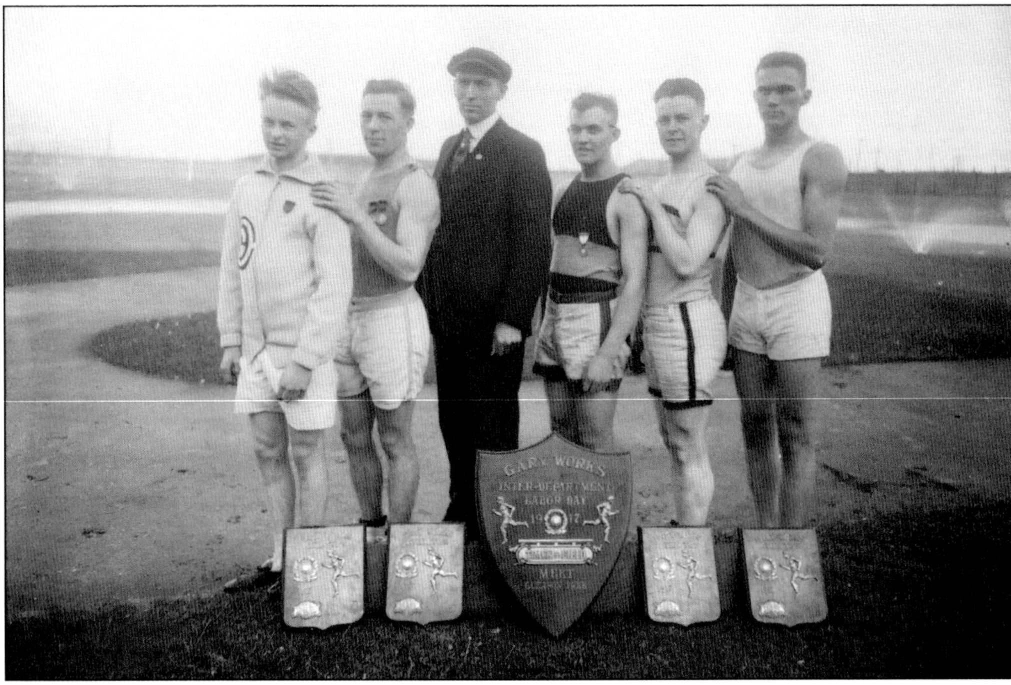

Mill Track Team. Track and field events were held at Gleason Field, and the departments competed against each other in friendly competition. This early 1920s photograph shows members of the chemical lab team with their awards. (Courtesy of CRA.)

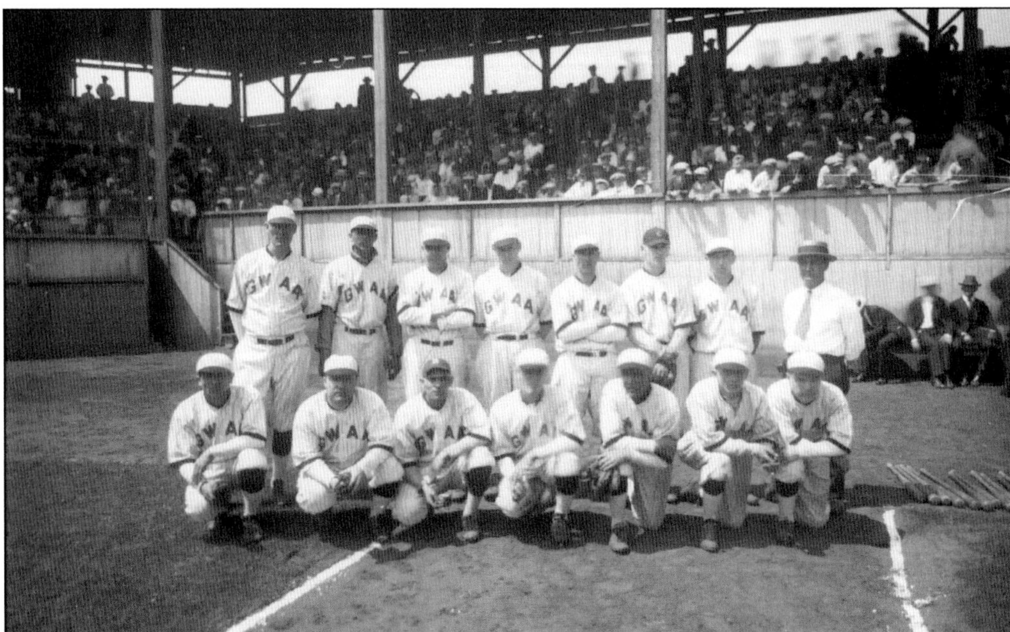

US Steel Baseball, 1926. US Steel sponsored a number of sports teams that were open to every department to encourage friendly competition among the workers. Every department, such as the coke plant, blast furnace, and open hearth, fielded teams in baseball, basketball, and track. A team is shown here at Gleason Field in 1926. (Courtesy of CRA.)

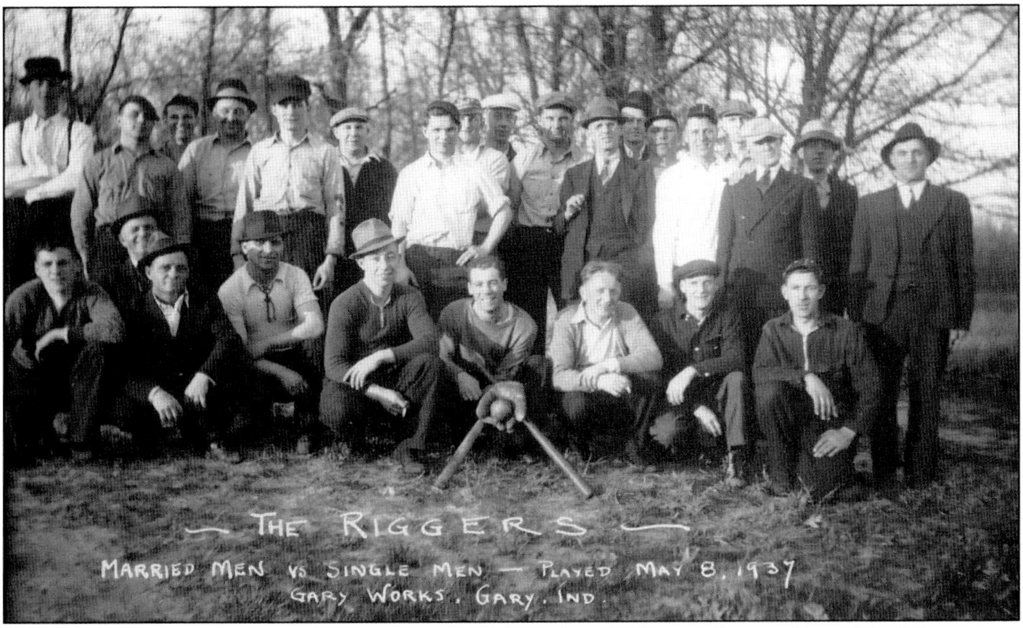

RIGGERS BASEBALL TEAM. The Riggers also fielded a baseball team during the 1930s. These were the men who repaired and changed the heavy cables that were used for the overhead cranes, the skips that pulled the loads of coal and ore to the top of the blast furnaces, and those used for the huge bridges. They also checked the cables that were used to load and unload trucks and railcars. They were a tough crew of men at US Steel's Gary plant. (Courtesy of CRA.)

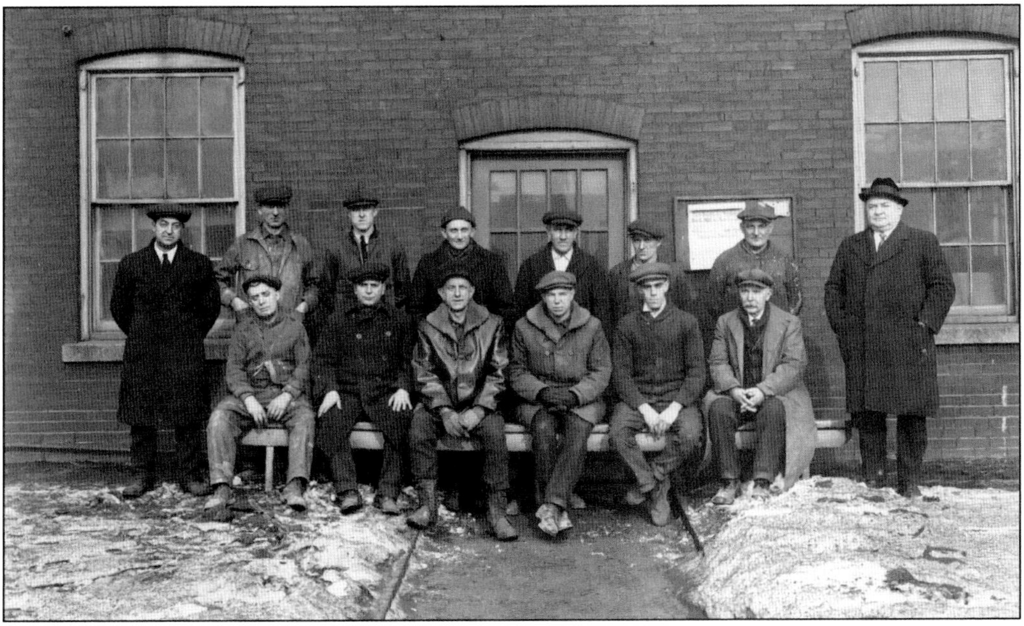

SAFETY COMMITTEE. To help curtail the number of on-the-job injuries that resulted in lost time, US Steel's Gary Works created a safety committee in the 1920s. In each department, committee members checked on workers to see if they were obeying the safety rules and also investigated serious accidents that made employees loose work time. Pictured is the safety committee in the 1920s. (Courtesy of CRA.)

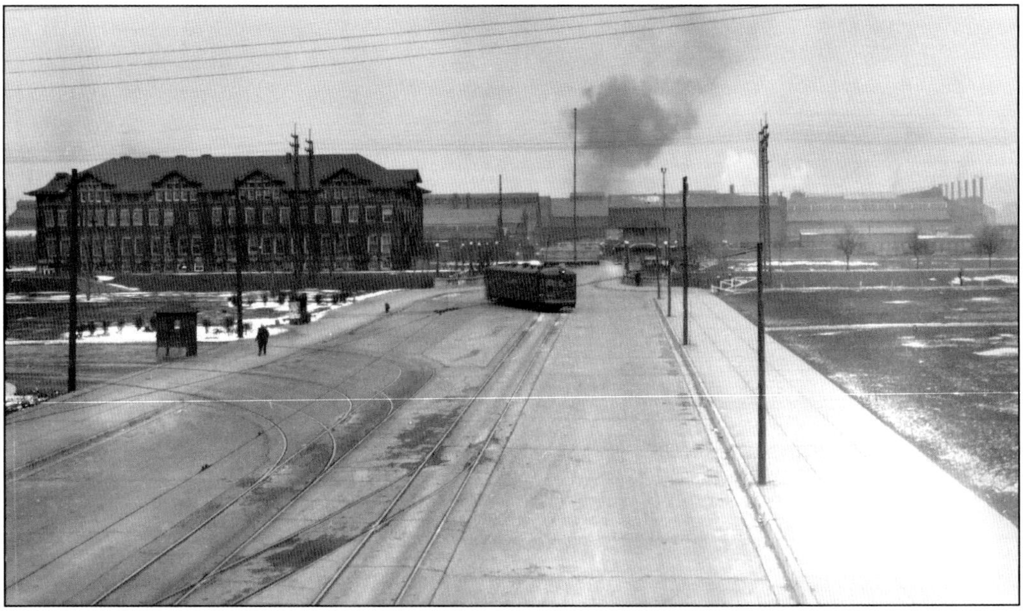

Main Gate, 1931. During the Great Depression, US Steel Corporation, like other steelmakers in the United States, went through a severe economic downturn. Orders were down, and the company stock prices fell. In 1931, the worst year of the Depression, the Gary plant operated only two of its twelve blast furnaces. The rest of the operations were at less than half capacity, and that hurt the city's downtown stores, causing further layoffs. Pictured is the Broadway main gate with few people coming or going. (Courtesy of CRA.)

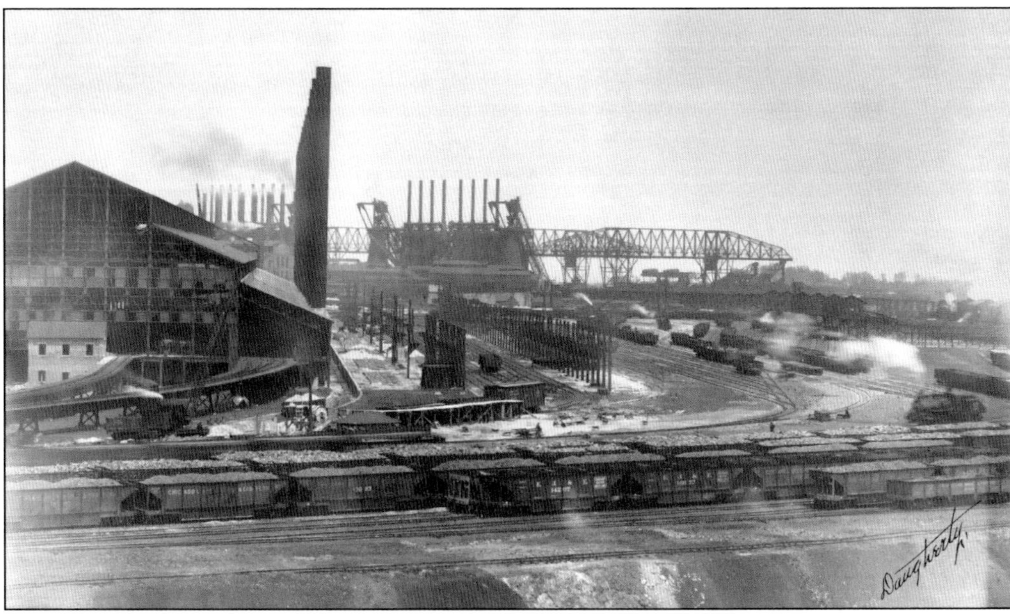

US Steel Preparing for War. As war clouds hung over Europe and the world in 1940, the United States began a defensive spending plan to be ready for possible war. US Steel, already under contract with the union, agreed to a new contract in April 1941. The company wanted to be able to meet steel orders from defense contractors and Great Britain without a threat of strike. In exchange for a raise, the workers agreed to a no-strike contract. (Courtesy of CRA.)

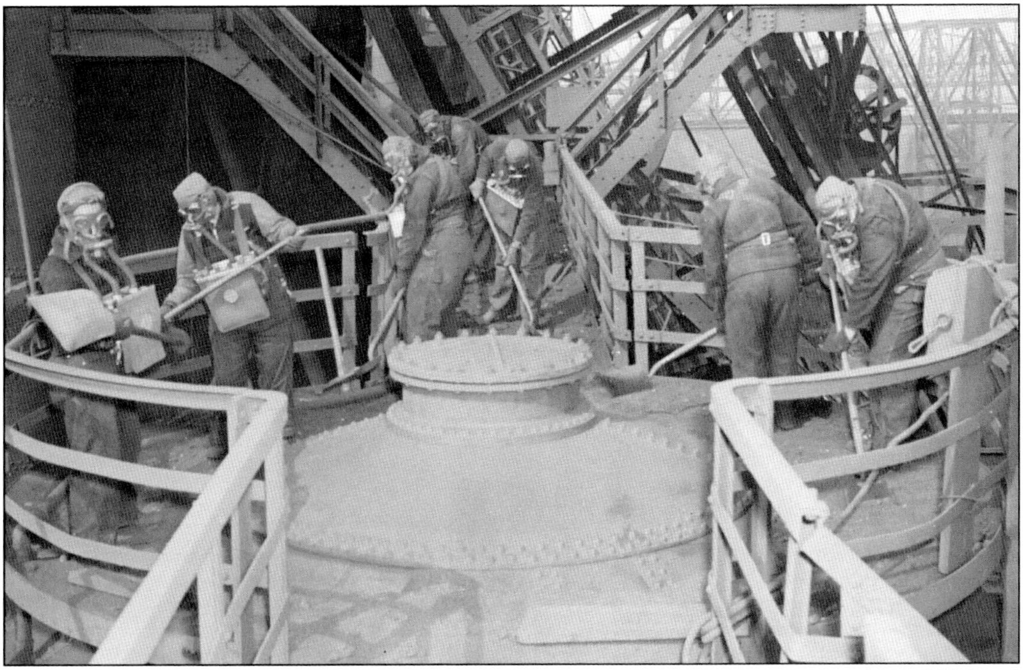

Women and Blast Furnace Work, 1943. Women took on the dangerous tasks at US Steel just as their male coworkers did. Here, they are working on top of the stoves next to No. 5 Blast Furnace. They are wearing gas masks to protect them from carbon monoxide fumes that are often present during operations near the furnaces and stove areas. They are working about six stories above ground. The nation owed a debt of sincere gratitude to these hardworking and patriotic American women. (Courtesy of CRA.)

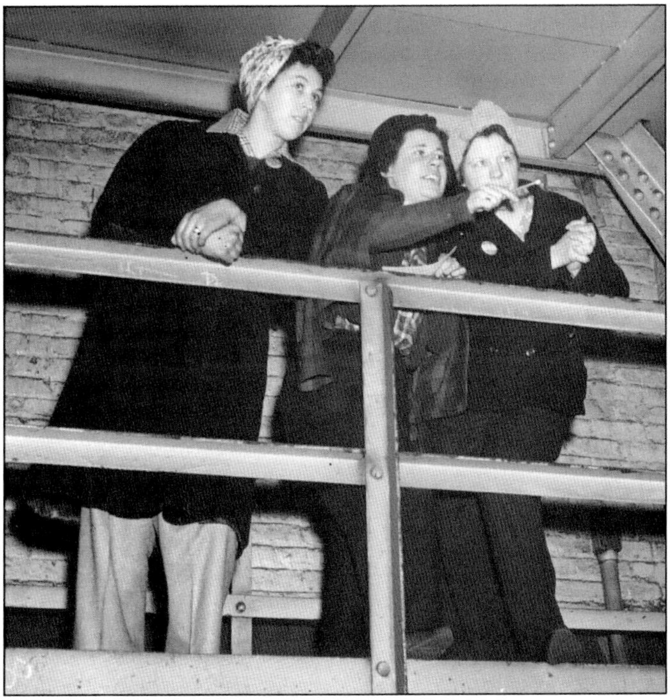

Women Employees, 1943. With 16 million men serving in the US military during World War II, American industries turned to the nation's women to keep production going. Mills, factories, ship and aircraft builders, railroads, and trucking firms hired women, known as "Rosie the Riveters." Other businesses also saw an increase in the number of women hired. The "Rosies" wanted to contribute to the war effort and bring victory and their loved ones home. Here, three Rosies are pictured in training in the Merchant Mill. Over 4,000 women worked in the Gary plant. (Courtesy of CRA.)

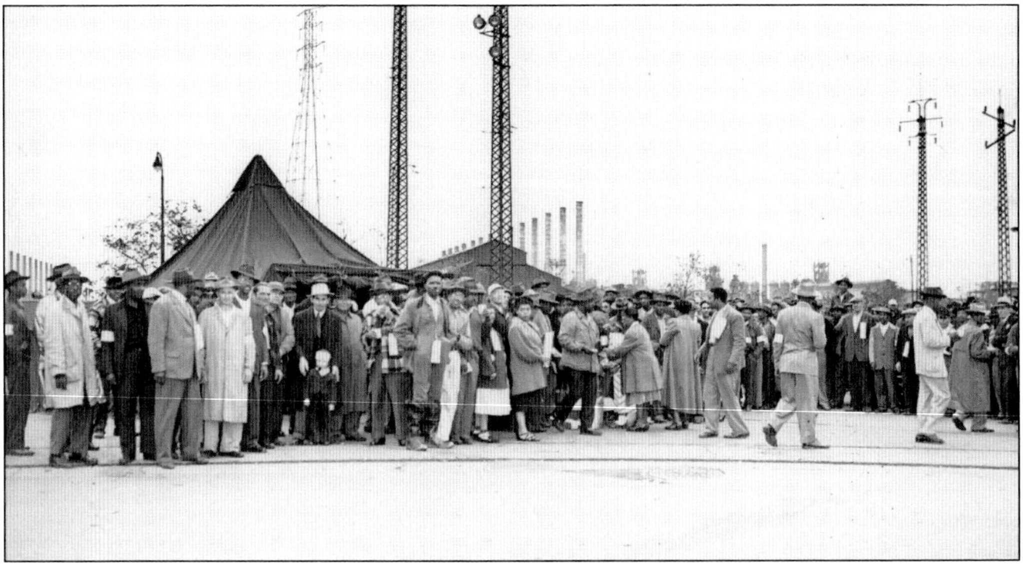

1949 Strike Pickets. In 1949, the membership of the United Steelworkers of America voted to strike against US Steel over pension and insurance benefits. After 42 days, union and company officials agreed to a new contract that brought about a noncontributory pension and sickness and life insurance coverage. Strikers are shown here at the Broadway gate. (Courtesy of CRA.)

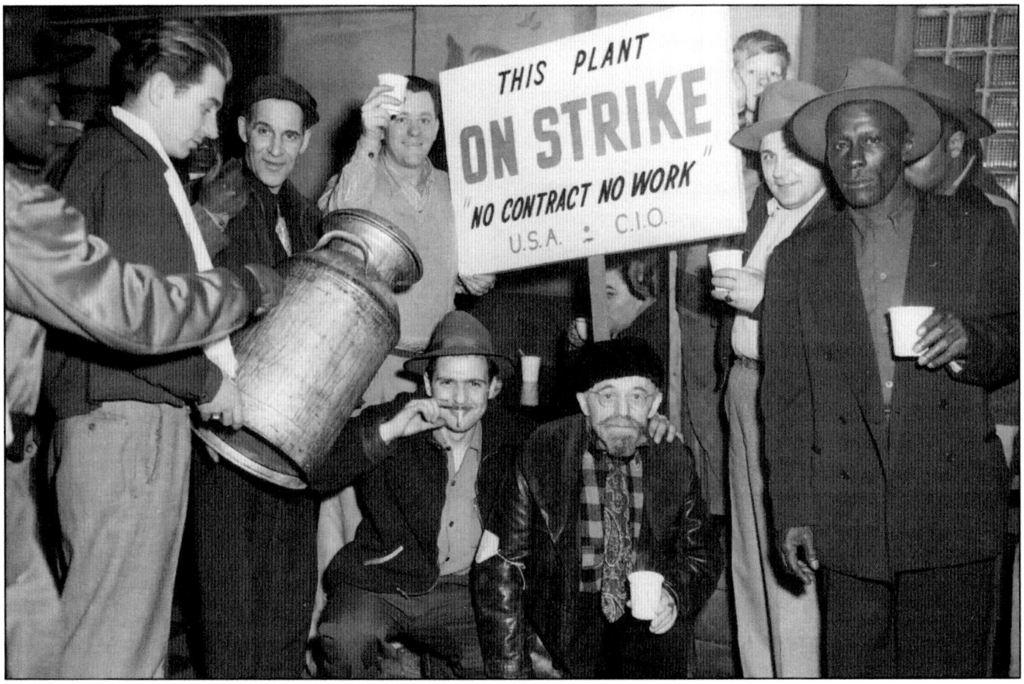

1952 Strike. Three years later, when the contract expired, the United Steelworkers walked out when management rejected a compromise and raise of 26¢ per hour. President Truman ordered the seizure of the mills as the shutdown was a national security threat that would hurt the Korean War effort. A 55-day walkout began. In July 1952, Truman ordered both sides to Washington, where a settlement was reached. Pictured are members of Local No. 1014 at their union hall in the Philip Murry Building at Fifth Avenue and Massachusetts Street. (Courtesy of CRA.)

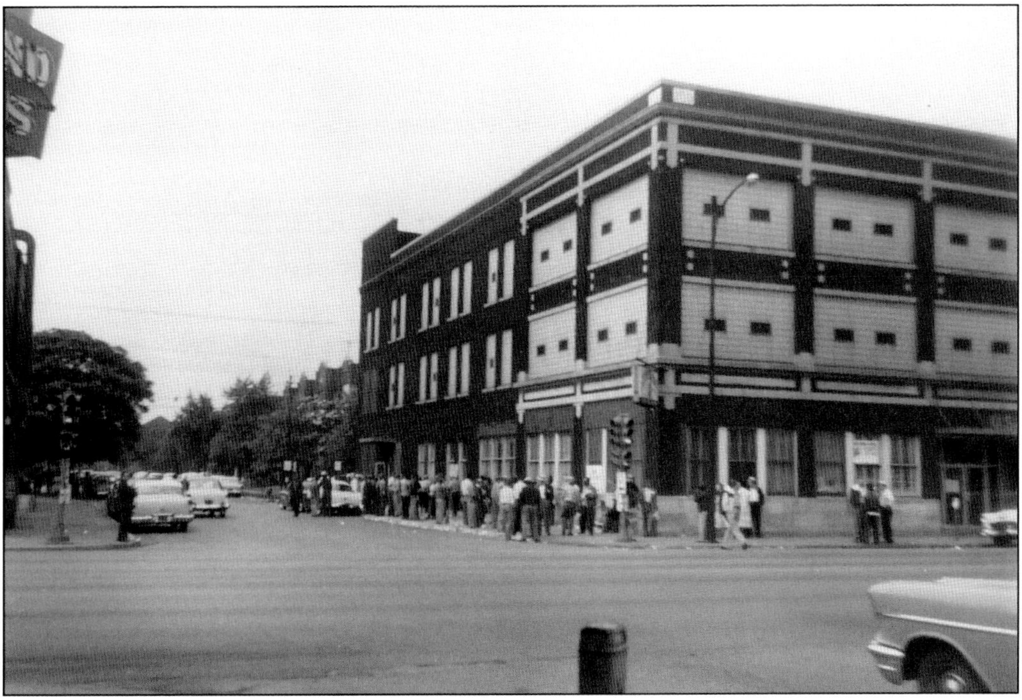

Philip Murray Hall, 1949. In the late 1940s, the United Steelworkers of America raised funds and built a new union hall on the northeast corner of Fifth Avenue and Massachusetts Street in Gary's downtown. The new facility featured new offices, meeting rooms, a cafeteria, and kitchen facilities, as well as a large meeting hall. The facility was used until the 1970s, when Local No. 1014 moved to McBride Hall on Texas Street near Interstate 65. (Courtesy of CRA.)

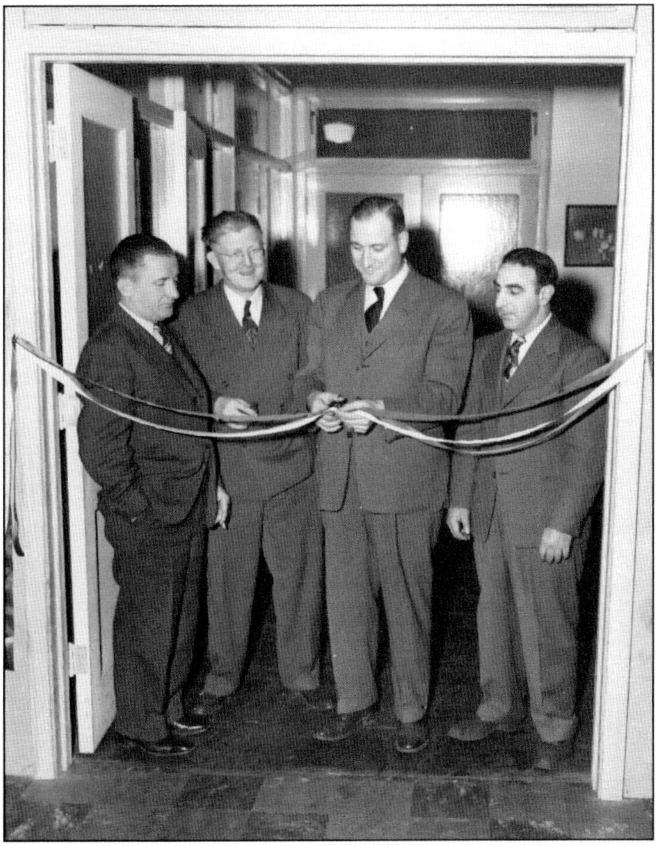

Ribbon Cutting, 1949. A formal ribbon-cutting ceremony was held for the local press, members of Local No. 1014, and families of the steelworkers. After the opening events, visitors were invited to tour the new union hall. (Courtesy of CRA.)

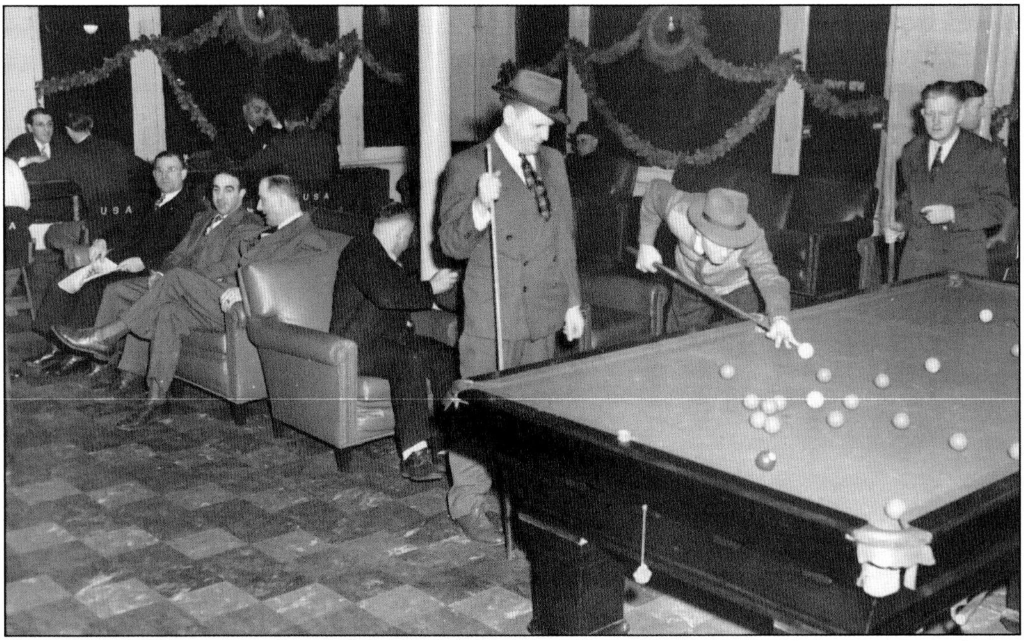

Murray Hall Lounge, 1949. The Phillip Murray Union Hall included a lounge. Pictured are a number of Local No. 1014 members playing a game of pool. Note the double-breasted suits, classic hats, and the man smoking at right. This was, of course, long before smoke-free buildings. Many of the members were probably World War II veterans who enjoyed smoking Camels or Lucky Strike cigarettes. (Courtesy of CRA.)

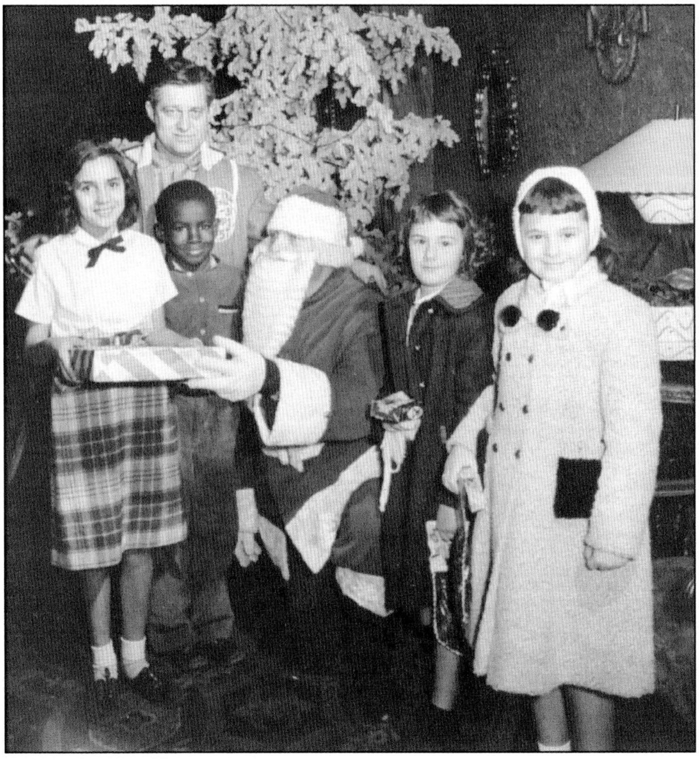

United Steelworkers of America Christmas, 1957. Like many groups and organizations, United Steelworkers of America District 31 was involved in the community as a good neighbor. At Christmas, union men and women looked out for those families that had fallen on difficult times. In addition, children of the steelworkers were invited to take part in the annual Christmas party. Pictured are children of union members in 1957. Of course, Santa was there at the union hall to pass out gifts. (Courtesy of CRA.)

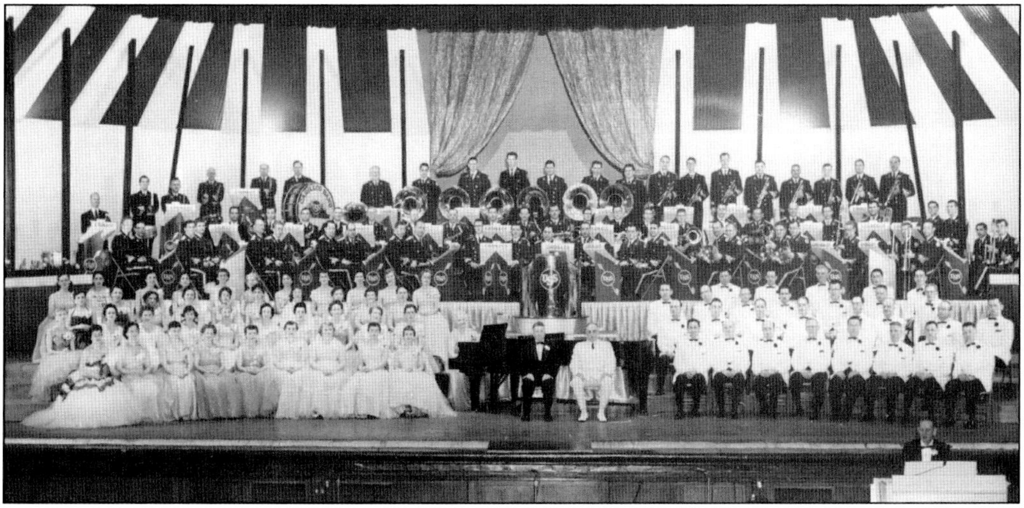

US Steel Band. The company sponsored a number of civic and cultural organizations for over 50 years in the Steel City. In addition, US Steel raised money for charitable and religious groups that assisted the less fortunate in Gary. Pictured is the US Steel band in 1957. Membership was open to all employees, staff, and administration. All that was needed was musical talent. (Courtesy of CRA.)

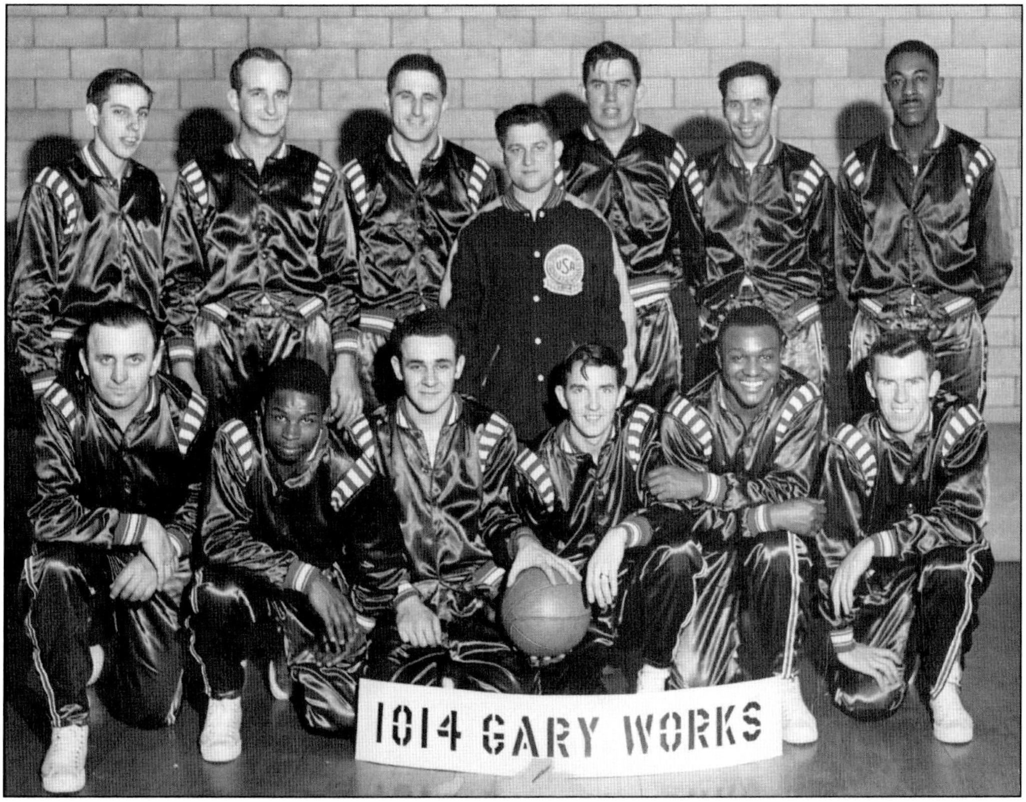

Steel Union Basketball Team, 1950s. The union sponsored athletic teams in the post–World War II years until the early 1970s. Pictured are members of the Local No. 1014 Gary Works basketball team in the 1950s. (Courtesy of CRA.)

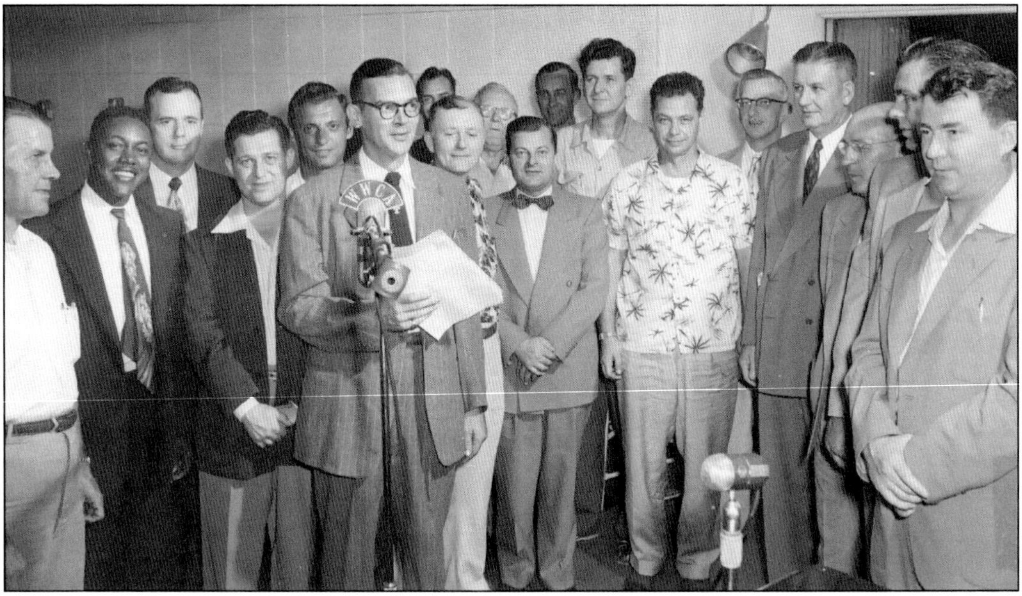

Union Meeting on Radio, 1950. In the 1950s, nearly 40 percent of American workers were members of a union. In Gary, the vast majority were members of unions, mostly in steel. They had fought hard in the 1930s to gain union recognition and made big gains getting contracts with major steel producers. In 1937, US Steel recognized the United Steelworkers of America, and other companies fell in line. Though television was in its infancy, radio provided shows geared to a blue-collar audience. WWCA Radio in Gary aired shows that dealt with issues important to steelworkers, their families, and the local community. Here, the radio host poses questions for members of Local No. 1014 in the studio. (Courtesy of CRA.)

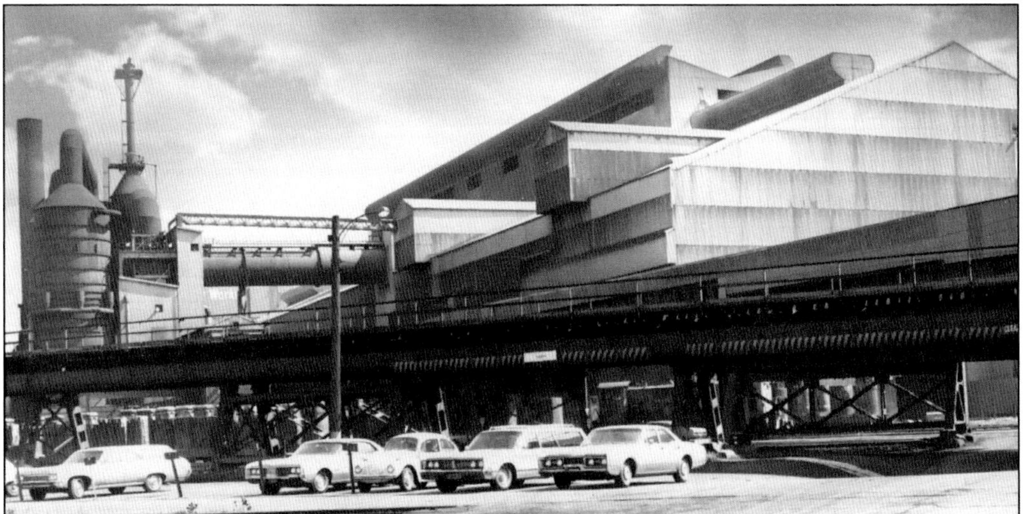

BOP Shop at Gary Works, 1967. In the 1960s, US Steel's Gary Works began to phase out the old open-hearth furnaces. The process resulted in better steel quality and increased profits. In addition, far less pollution was released into the air. Years ago, travelers driving north toward Northwest Indiana knew that they were getting closer to Gary and the surrounding area by the red dust in the sky. Pictured here is the No. 1 BOP Shop about a half mile west of the blast furnaces. (Courtesy of CRA.)

Three

Politics and City Government

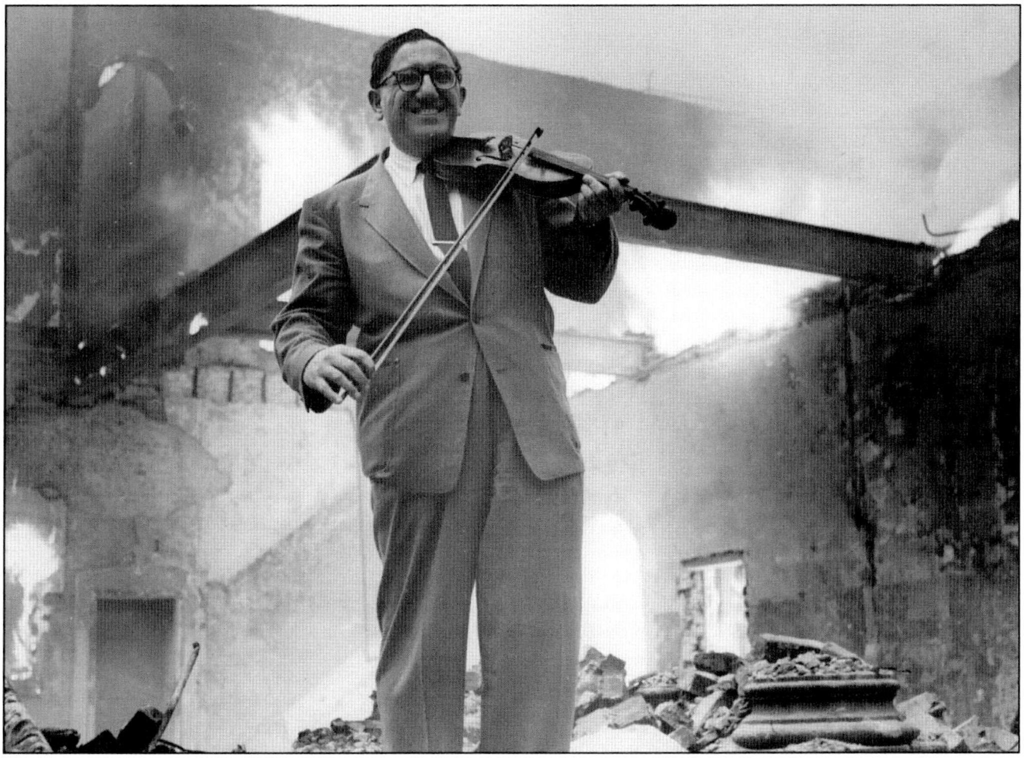

Chacharis with Violin, 1950s. Politics was always part of life downtown, starting with the 1908 election of Thomas Knotts, Gary's first mayor. Political issues would determine the direction of Gary's growth, relations with US Steel, the union movement of the 1930s, and civil rights. The Steel City started with a Democrat in the mayor's chair and then a Republican until the early 1930s, when the Great Depression took hold. Since then, Gary has been run by Democrats, except for one term of Ernest Schaible in 1939. The rise of the United Steelworkers of America was a big factor. Gary had its colorful and controversial mayors as well. Roswell O. Johnson served three separate terms as mayor, his last after serving time in federal prison for violating the Volstead Act. His popularity did not help him gain a fourth term, as Republicans were not successful during the 1930s. George Chacharis, a popular Democrat, also spent time in prison over tax issues and federal corruption charges. Many voters believed that if he could legally run, Chacharis might have won again. He is pictured here playing a violin amid the rubble of the first city hall in the 1950s. (Courtesy of CRA.)

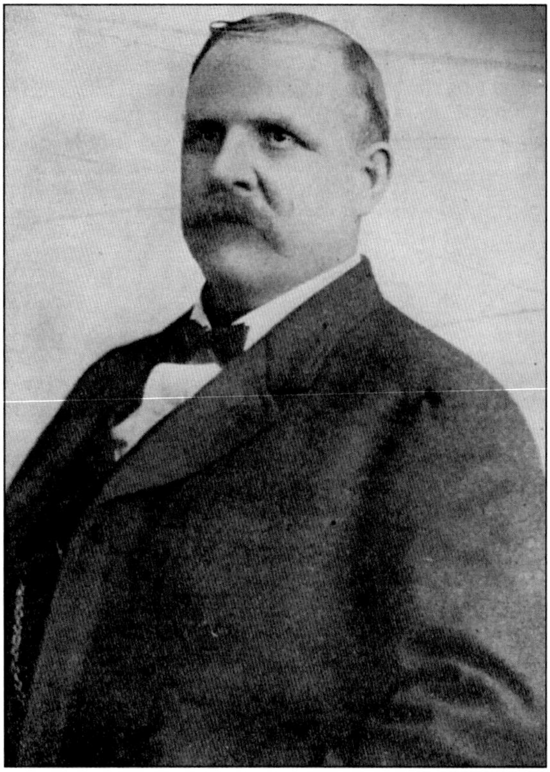

THOMAS KNOTTS, 1908. Gary's first chief executive, Thomas E. Knotts, came to the area when the site for US Steel was selected. In the town's first election in 1909, Knotts defeated the company's candidate, John A. Brennan, in what turned out to be a wild affair of name-calling, threats, and accusations of dirty politics. Knotts was at odds with US Steel throughout his only term but did gain success in the young city's expansion. Tolleston and Glen Park were incorporated. Thomas Knotts would be the Steel City's only Democrat to lead Gary until the 1930s. The stock market crash of 1929, Great Depression, and the New Deal changed the politics of Gary and Lake County from Republican to Democrat. (Courtesy of CRA.)

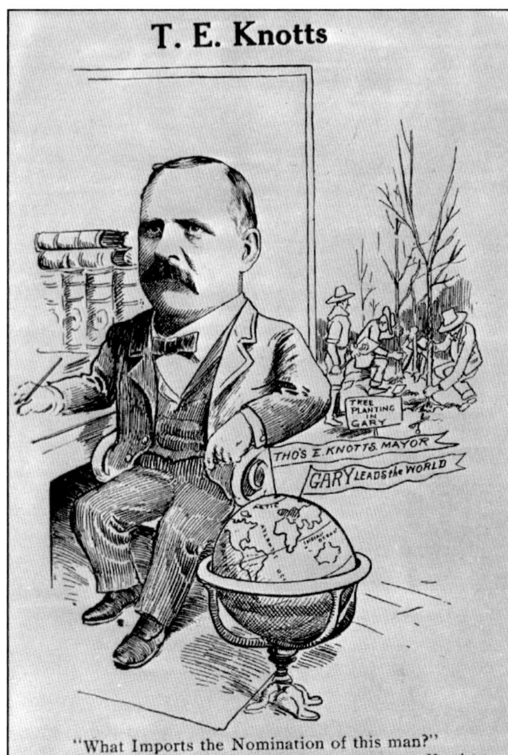

TOM KNOTTS EDITORIAL. Like all politicians, Tom Knotts faced critics and supporters in the local media of his day. Pictured is Knotts portrayed in an editorial cartoon in 1910. (Courtesy of CRA.)

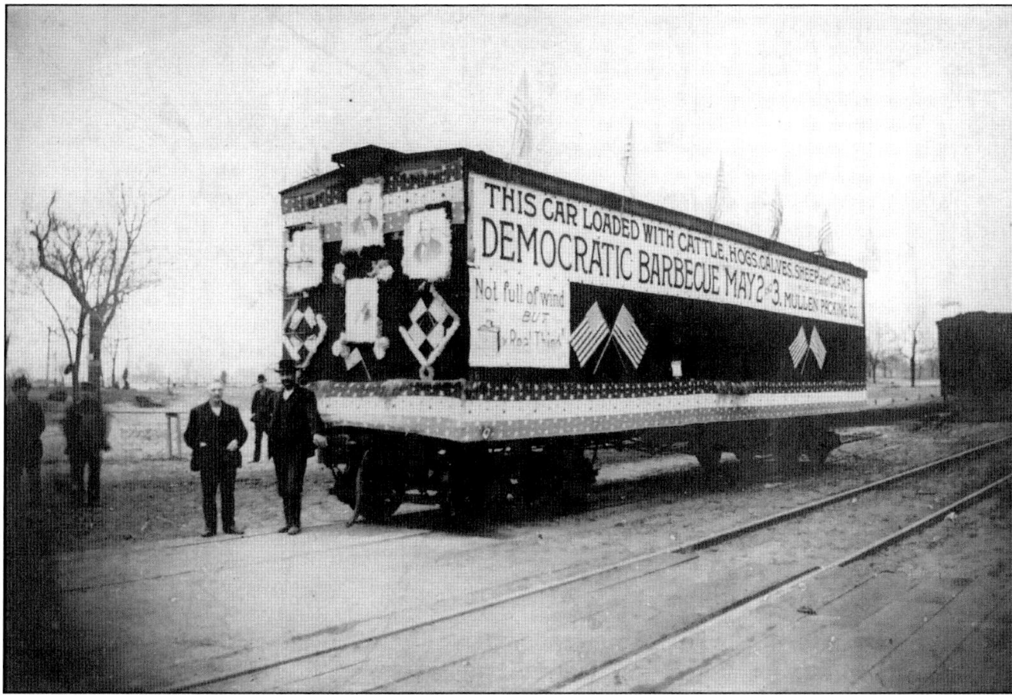

CAMPAIGN TRAIN, 1908. All methods were used to increase voter turnout in the young city in 1908. Here, residents were promised a barbecue if they would come out to a rally, meet candidates, eat, and of course, vote for the "best man." Not much has changed in local elections. (Courtesy of CRA.)

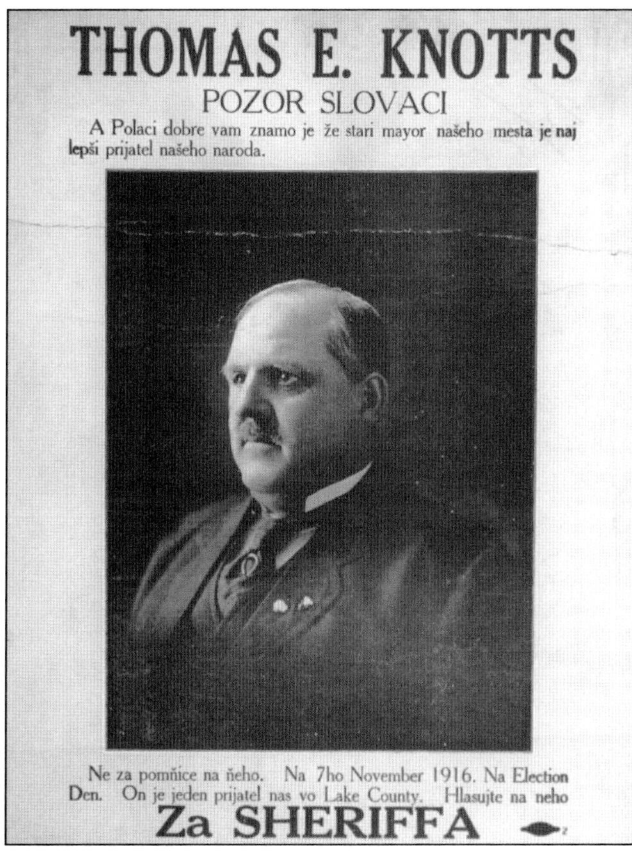

KNOTTS ELECTION POSTER, 1909. In 1916, the former mayor threw his hat in the ring as a candidate for county sheriff. The former mayor of the Steel City had support in the immigrant communities and his campaign staff put out posters to appeal to the Slavic population in the county. (Courtesy of CRA.)

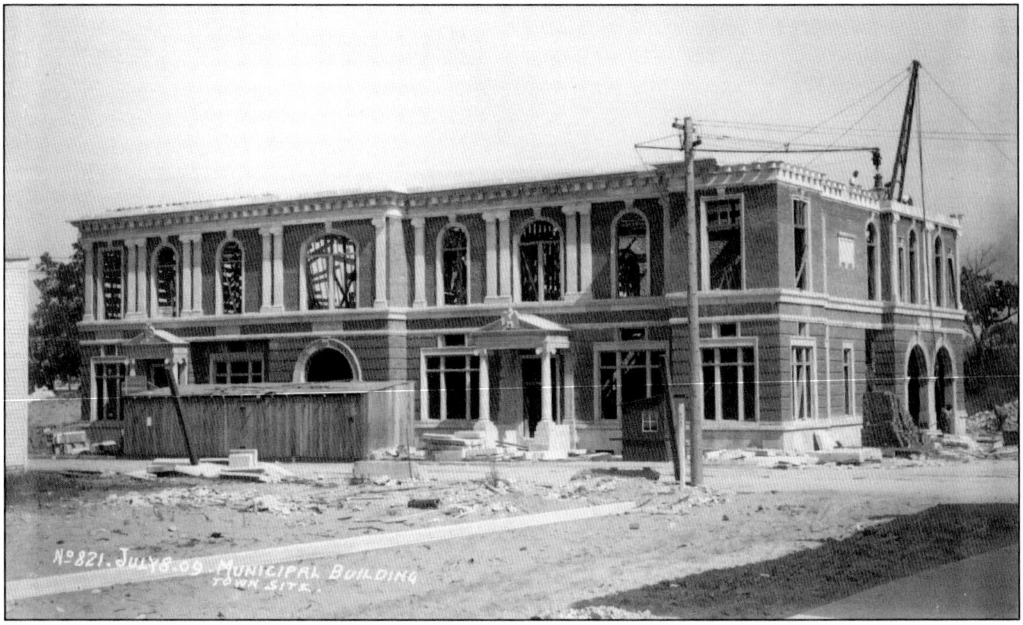

FIRST MUNICIPAL BUILDING, 1909. In 1909, the first municipal building was under construction at the northwest corner of Seventh Avenue and Massachusetts Street. It contained all city offices, council chambers, and the downtown police and fire stations. It sat just across the street from Memorial Auditorium until the early 1950s. In the 1920s, the mayor's office and council chambers moved to the new city hall at Fourth Avenue and Broadway. The police station and city jail were moved to the new municipal building at Thirteenth Avenue and Broadway in 1955. Soon after, the old building was razed and replaced by a carport. (Courtesy of CRA.)

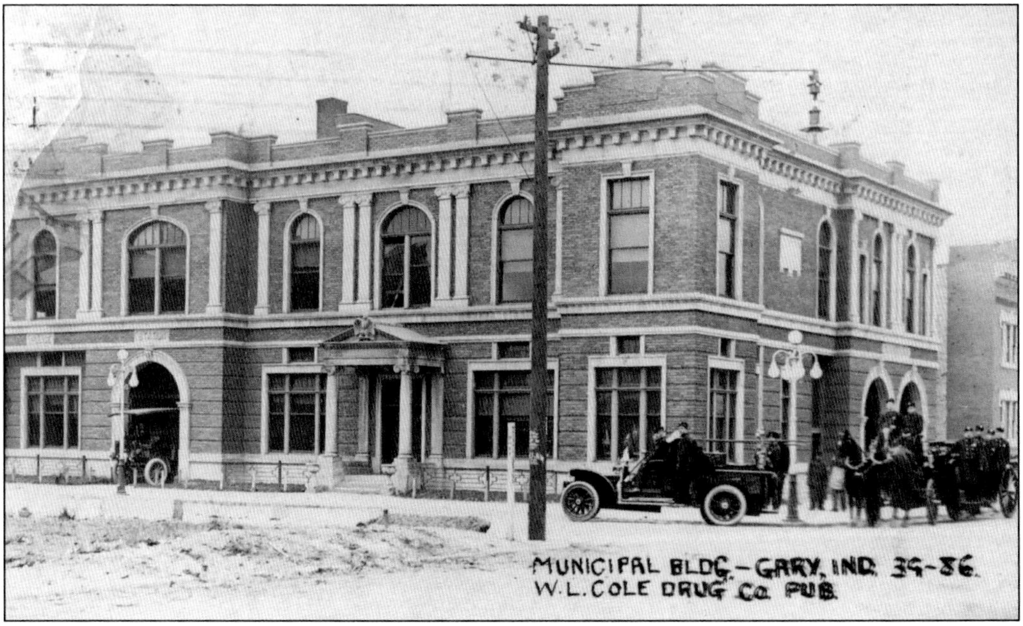

FIRST CITY HALL, 1907. Gary's first city hall was located at Seventh Avenue and Massachusetts Street and would be home to the police station, main firehouse, and all the city's municipal offices. It is pictured here under construction in July 1907. (Courtesy of CRA.)

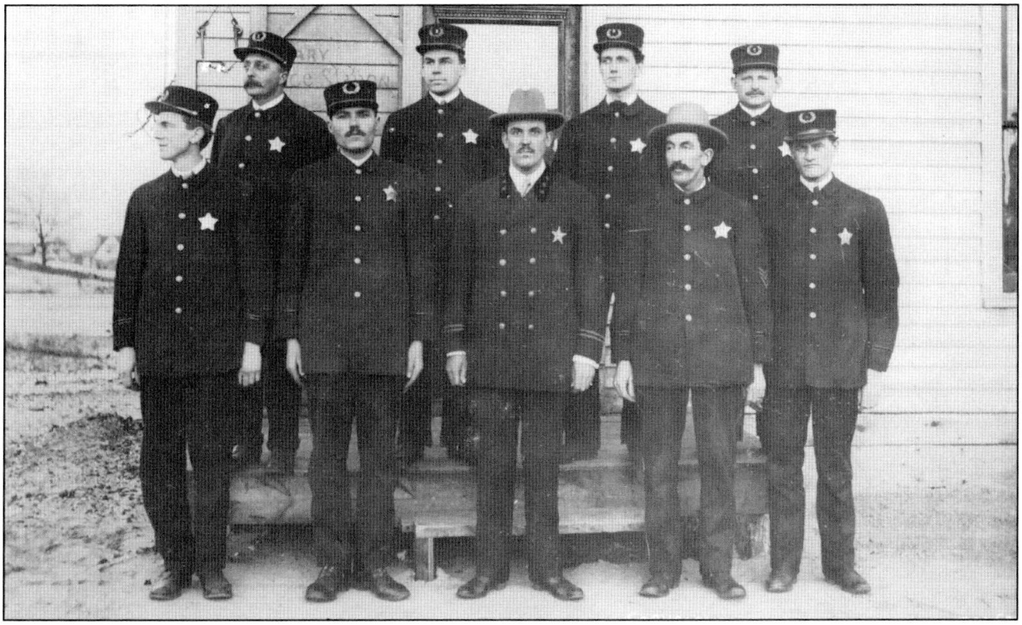

GARY POLICE, 1908. In the early years of the Steel City's history, Gary was often looked upon as a Wild West town with crime out of control in certain areas. Company police enforced the law on plant property and the area around the Grand Calumet River. As the city's population began to grow and residential areas expanded, the local government increased the number of police officers. The city force is pictured here in 1908. The group looks as if they mean business. (Courtesy of CRA.)

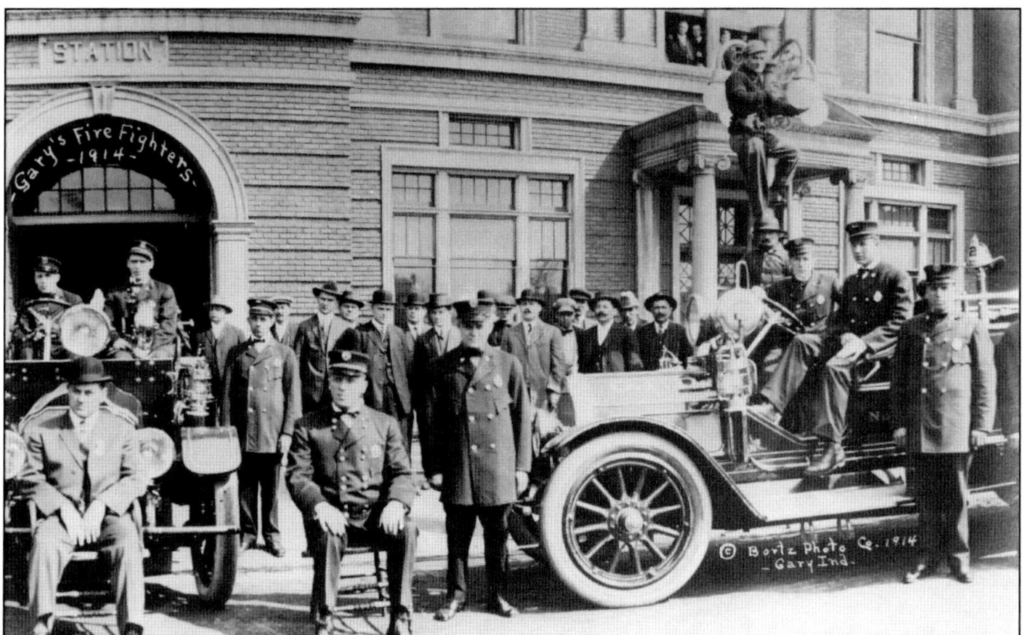

FIRE DEPARTMENT, 1914. Gary firemen pose in front of city hall in 1914 with their firefighting equipment. The Steel City's first fire department was a volunteer force run by the police department until 1908. Joseph J. Feely organized the professional fire force that year. (Courtesy of CRA.)

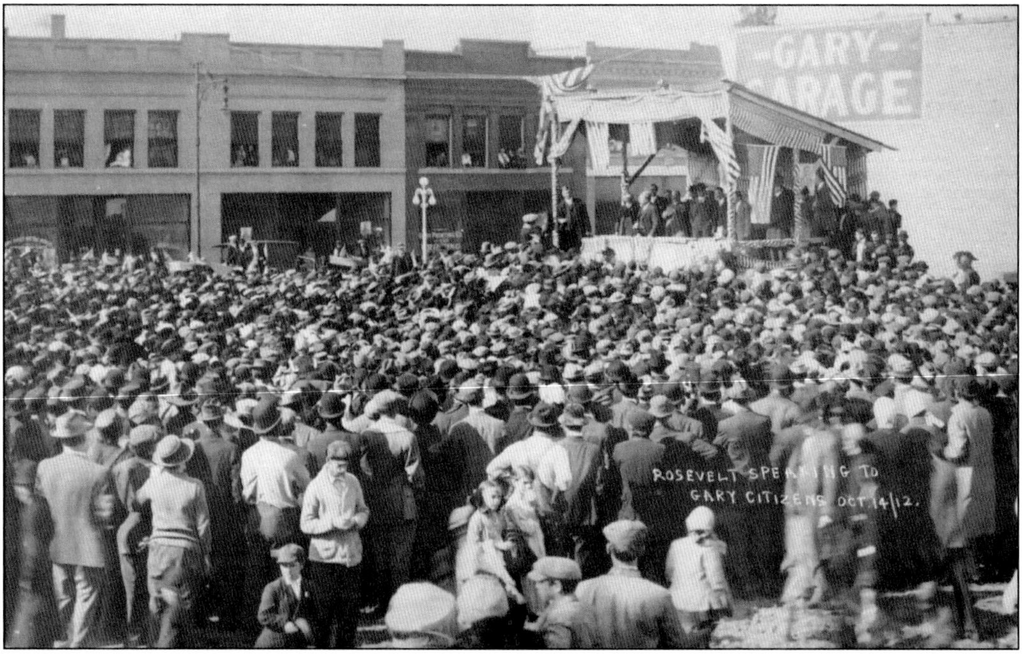

TEDDY ROOSEVELT IN GARY, 1912. In the presidential election of 1912, three of the four candidates came to campaign in the city of Gary. Democrat Woodrow Wilson, Socialist Eugene Debs, and Bull Moose Party choice Teddy Roosevelt all spoke downtown. Here, Roosevelt addresses the crowd on Broadway in October 1912. A few days after Roosevelt spoke in Gary, an assassin tried to shoot him in Milwaukee, Wisconsin. He was saved by papers in the inside pocket of his overcoat. (Courtesy of CRA.)

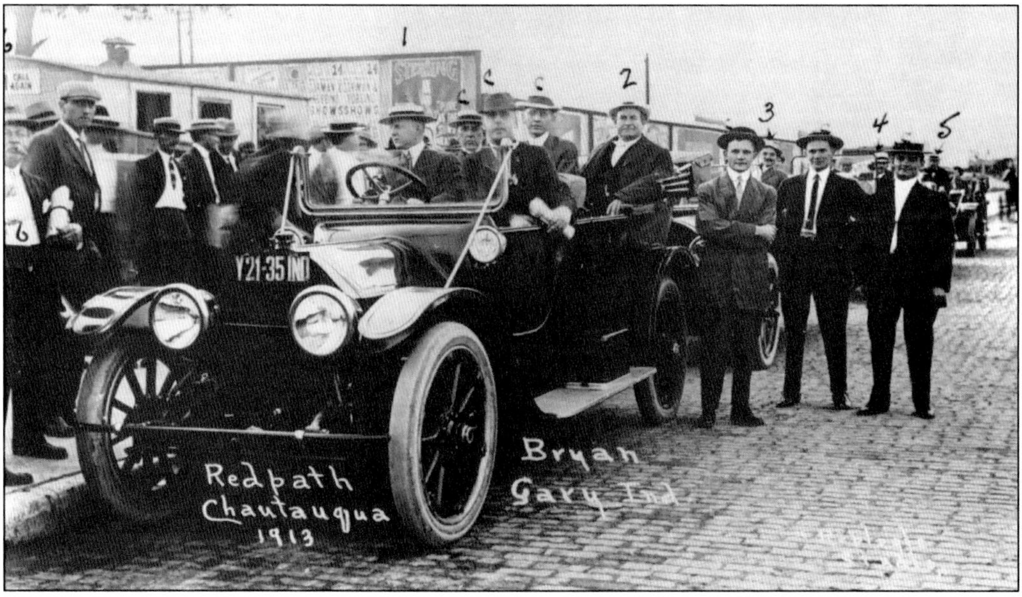

WILLIAM JENNINGS BRYAN, 1913. Former presidential candidate William Jennings Bryan paid a visit to the Steel City in 1913. At the time, he served as President Wilson's secretary of state. Prior to that, he ran for president in 1896, 1900, and 1908. In 1925, he was part of the famous Scopes Trial in Tennessee, working for the prosecution. (Courtesy of CRA.)

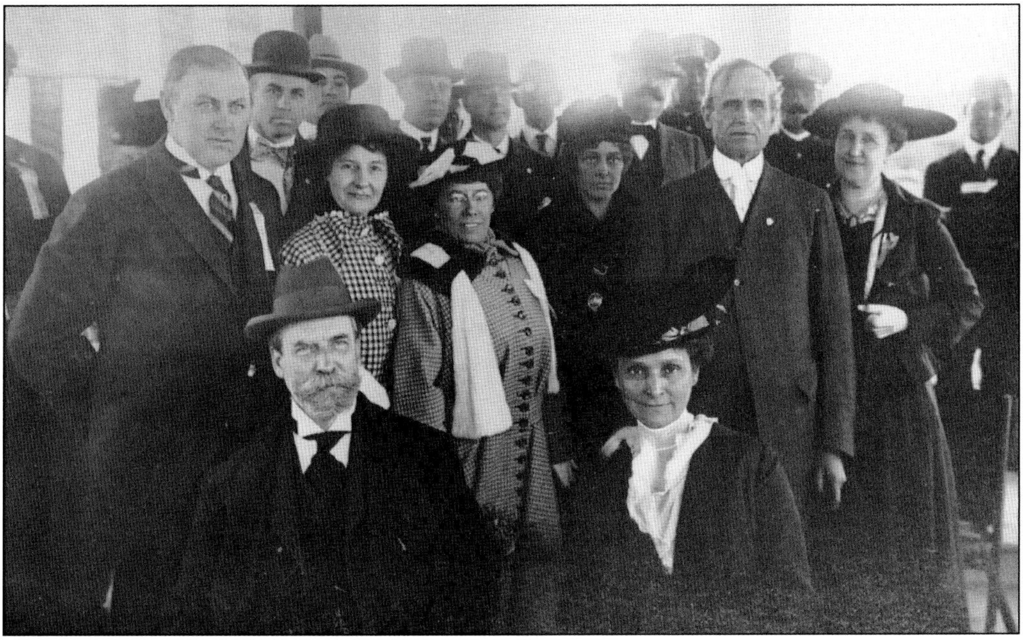

Charles Evans Hughes, 1916. In 1916, Charles Evans Hughes, an associate justice of the US Supreme Court, resigned to accept the Republican nomination for president of the United States. He lost to incumbent Woodrow Wilson, and in 1921 served as President Harding's secretary of state. Later, he was nominated to chief justice of the US Supreme Court, where he served for 11 years. He is pictured here campaigning in 1916 seated at far left with Gary mayor R.O. Johnson standing behind him. (Courtesy of CRA.)

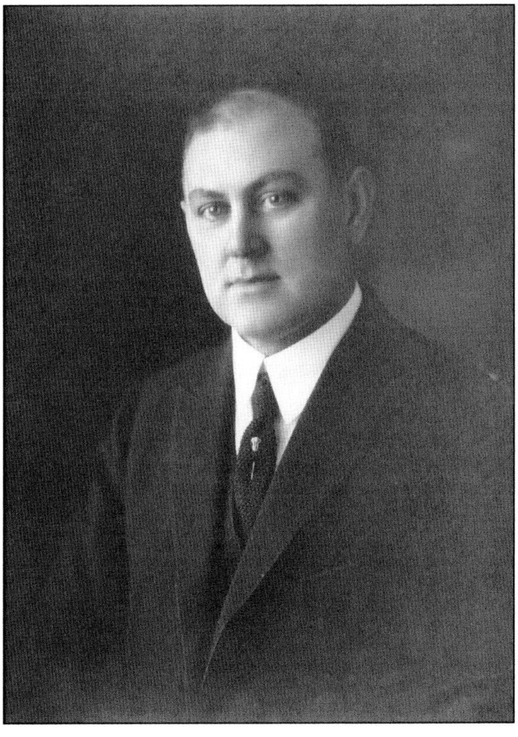

R.O. Johnson, 1913. Roswell O. Johnson, Gary's second mayor, won the election of 1913, defeating Thomas Knotts. Johnson, a three-time Republican mayor, was elected in 1913, 1921, and 1929. During the 1920s, he was found guilty of violating Prohibition laws and was sent to serve time in Atlanta. In 1929, President Hoover pardoned Johnson, and the former mayor ran again. He won a third time, but after the crash of 1929 and the Great Depression, Johnson was sent out of office by the voters. He and other Republicans were blamed by many voters for the economic disaster of the 1930s. (Courtesy of CRA.)

IF you ***WORK*** for a man,
in heaven's name, ***work for him***

If he pays you wages, that supply you bread and butter, work for him. Speak well of him. Stand by him, and the institution he represents. If put to a pinch, an ounce of loyalty is worth a pound of cleverness. If you must vilify, condemn or disparage, RESIGN YOUR POSITION. And when you are outside, condemn to your heart's content. But as long as you are a part of the institution, *do not condemn it.* If you do, you are loosening the tendrils that bind you to the institution. And the first high wind that comes along you will be uprooted and blown away in the blizzard's track, and you may never know WHY.

—— *Mayor R. O. Johnson.*

R.O. JOHNSON PLACARD, 1920s. Mayor Johnson expected his administration to carry out their assigned duties while they worked for the city. It may not be as famous as Harry Truman's "the buck stops here," but this motto did call on employees to give an honest day's work for their pay. (Courtesy of CRA.)

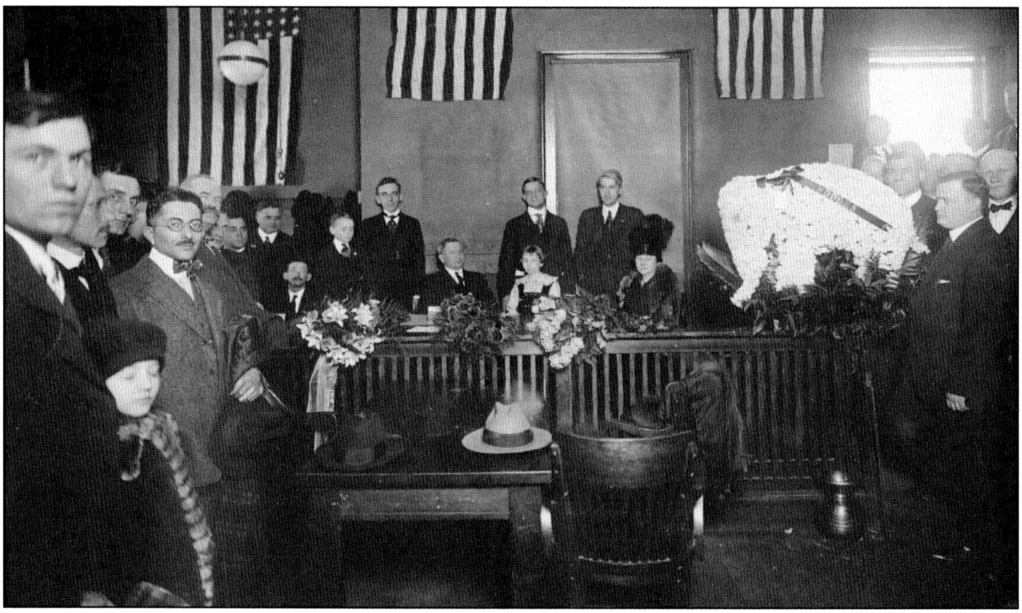

WILLIAM HODGES INAUGURATION, 1918. In 1917, Gary voters decided to replace Roswell Johnson with Republican William F. Hodges to serve as the city's mayor. Hodges sat at the chief executive's desk during the final year of the Great War of 1917–1918 and during the steel strike of 1919. Hodges is pictured here following his inauguration on January 7, 1918. (Courtesy of CRA.)

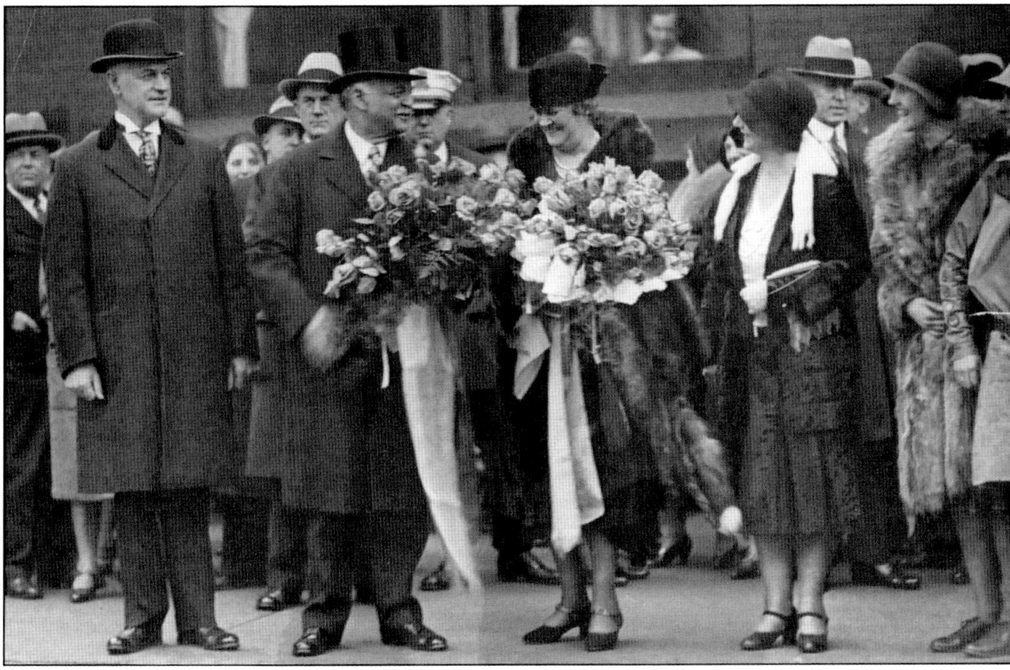

Charles Curtiss, Vice President, 1931. Even Vice Pres. Charles Curtiss came to Gary to rally Republicans for the 1932 election. However, the Herbert Hoover administration lost to Roosevelt in a landslide. Curtiss was the first Native American vice-president. (Courtesy of CRA.)

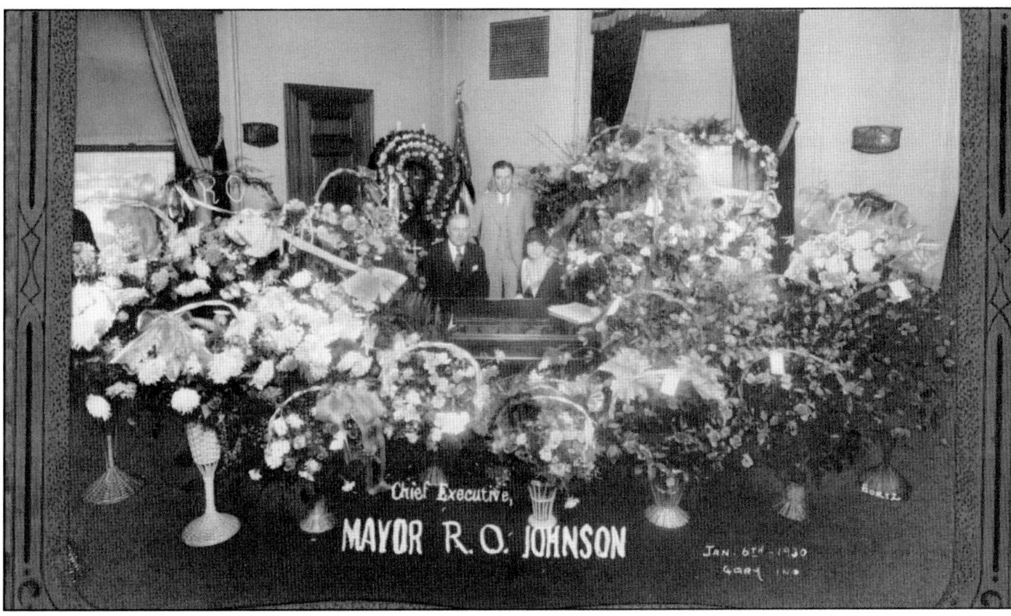

R.O. Johnson, 1930. After he had been all but written off as a mayoral candidate, R.O. Johnson won the election of 1929; the voters of the precincts south of the Wabash Railroad did not seem to care that he spent time in federal prison for violating the 18th Amendment. It would be the Depression and bad economic times for the Steel City that made those citizens vote out Johnson in the next election. Most would vote Democratic and abandon the Grand Old Party in Lake County. Johnson is pictured here in 1930. (Courtesy of CRA.)

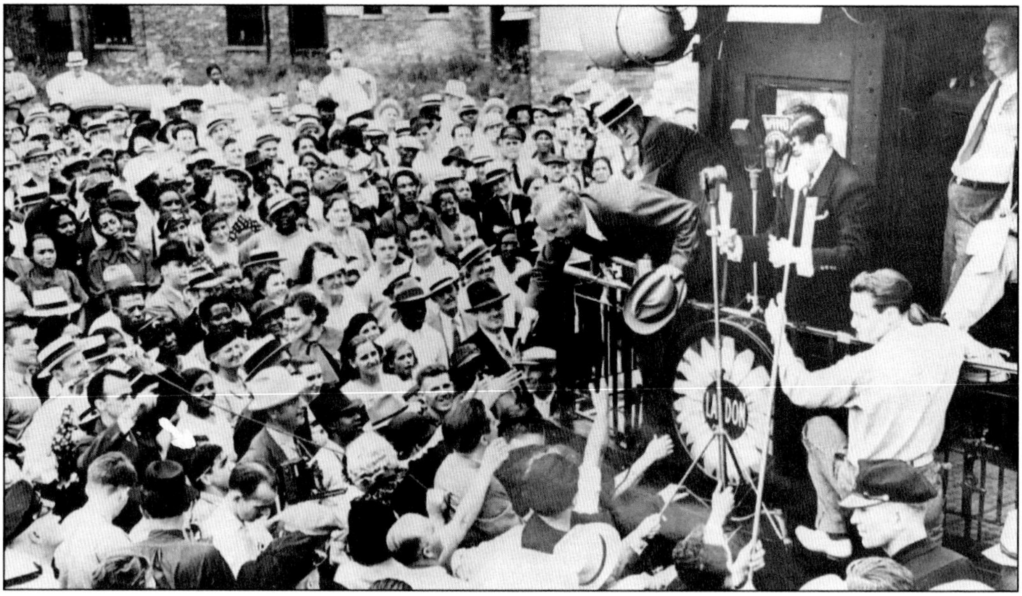

ALF LANDON CAMPAIGN, 1936. In 1936, Pres. Franklin Roosevelt sought a second term in the White House. The Republican Party nominated Alfred E. Landon to run against him. Landon is pictured shaking hands and campaigning during his whistle-stop in Gary. Landon, the progressive governor of Kansas, accepted the need of many New Deal programs but was critical of the costs and inefficiency associated with them. However, that did not matter, as Roosevelt was reelected in one of the biggest landslides in history. (Courtesy of CRA.)

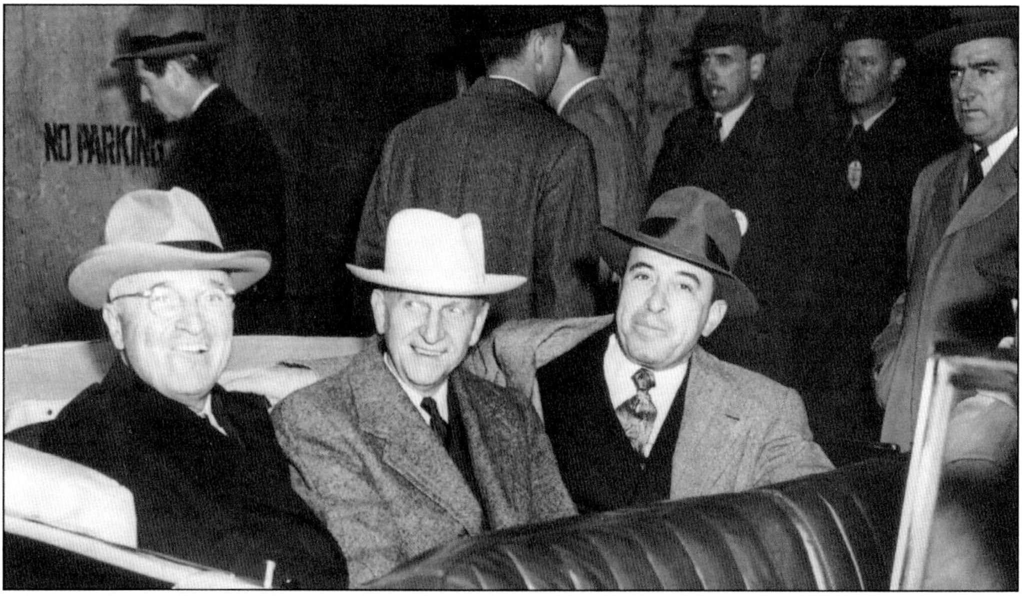

TRUMAN AND SWARTZ, 1948. In the 1948 presidential election, it looked as if Harry Truman would lose to the Republican challenger, Tom Dewey. The Democratic Party was split over the civil rights platform that caused a walkout by Southern Democrats at the convention. Truman did not give up and traveled by rail across the country to rally his voters. While on a stop in Gary, he gave a powerful speech at Memorial Auditorium, condemning the Republican Party and the "do-nothing Congress." He is pictured here with Gary mayor Eugene Swartz. (Courtesy of CRA.)

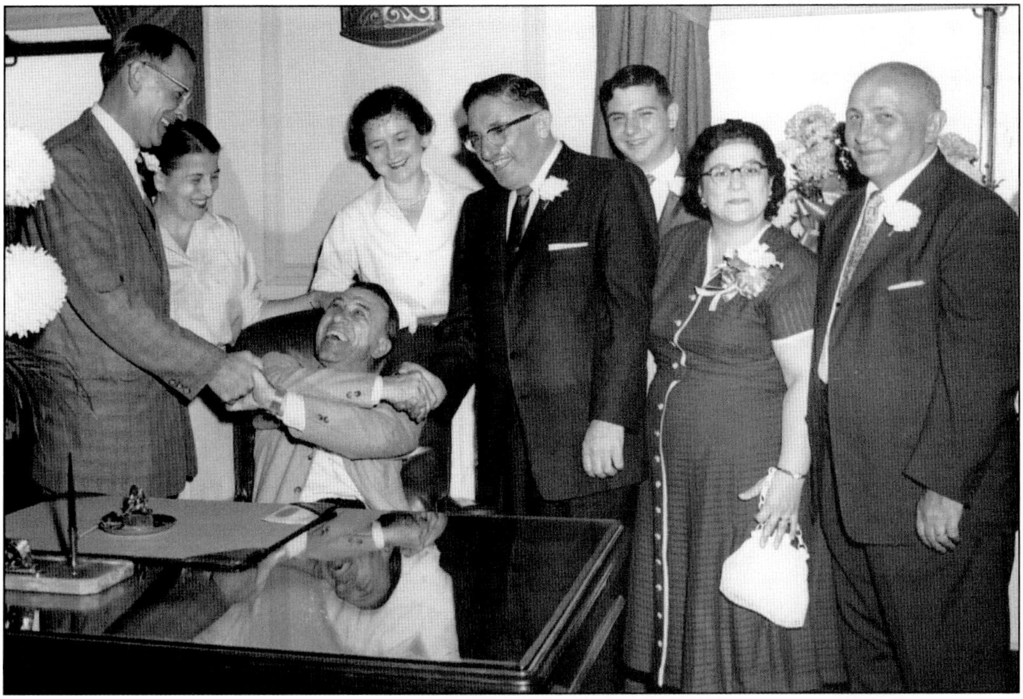

Passing the Torch, 1952. Pete Mandich was victorious in the November 1951 mayoral election against his GOP rival Paul Cyr. During the inauguration ceremonies of January 1952, Mandich, at left, is shown shaking hands with the outgoing mayor, Eugene H. Swartz. Shaking his other hand is George Chacharis, who was appointed city controller and later was elected mayor in 1959. (Courtesy of CRA.)

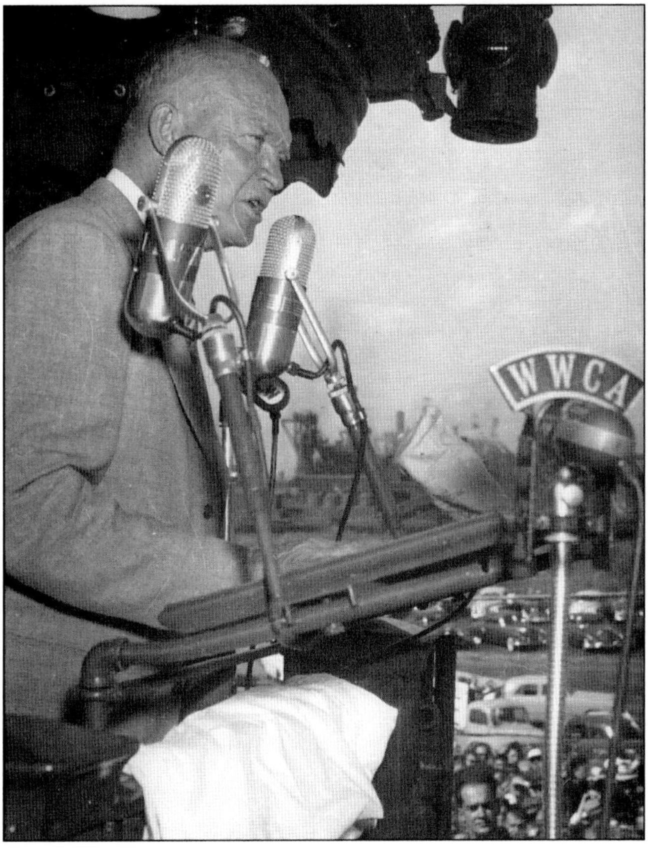

Eisenhower Campaigning, 1952. During the 1952 election, both major party candidates, Democrat Adlai Stevenson and Republican Dwight Eisenhower, made stops in the Steel City. Eisenhower was popular with World War II veterans, as well as many Democrats, and won two terms, 1952 and 1956. He is shown making a whistle-stop at the B&O station on Broadway. (Courtesy of CRA.)

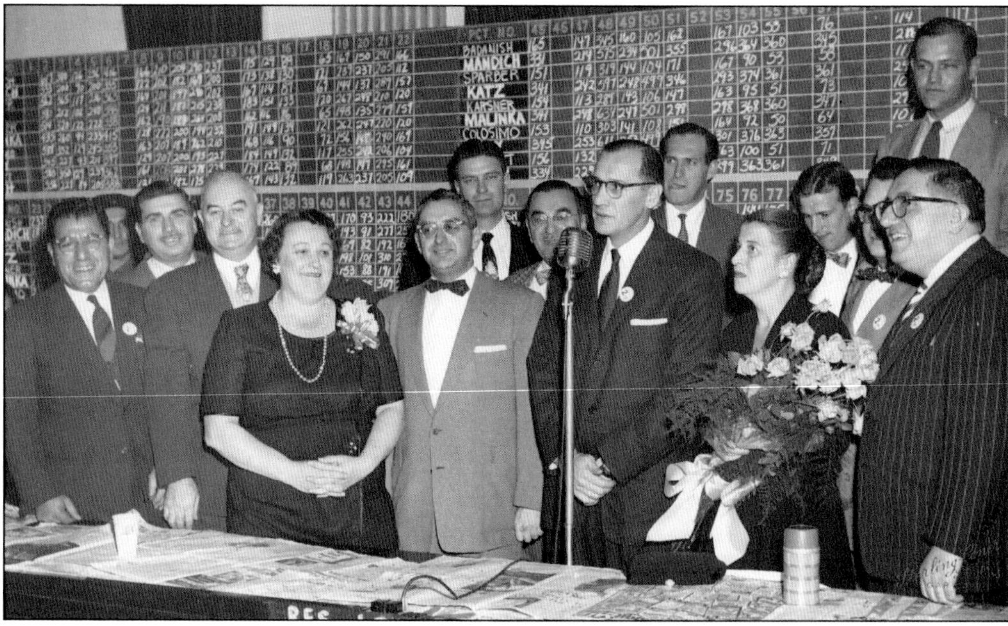

Election Night, 1955. Here, Mayor Pete Mandich and fellow Democrats await the results of the 1955 mayoral election. Mandich and his administration would win another term when the results came in. (Courtesy of CRA.)

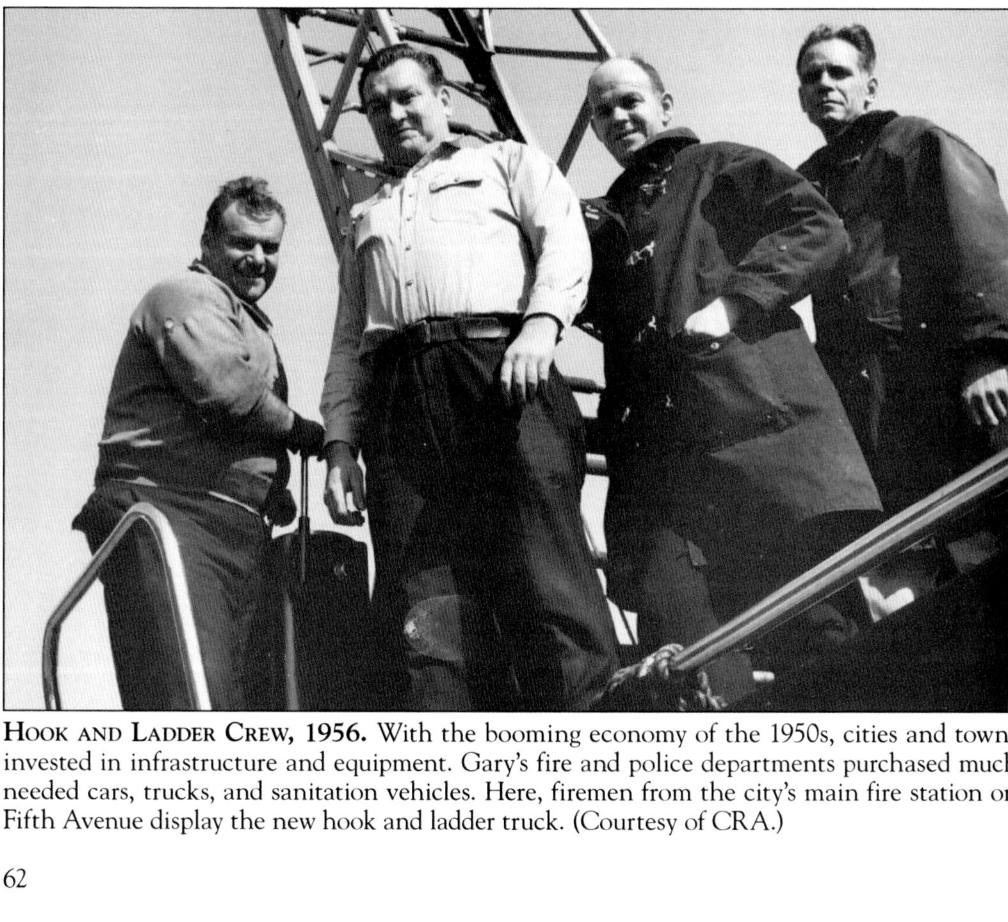

Hook and Ladder Crew, 1956. With the booming economy of the 1950s, cities and towns invested in infrastructure and equipment. Gary's fire and police departments purchased much needed cars, trucks, and sanitation vehicles. Here, firemen from the city's main fire station on Fifth Avenue display the new hook and ladder truck. (Courtesy of CRA.)

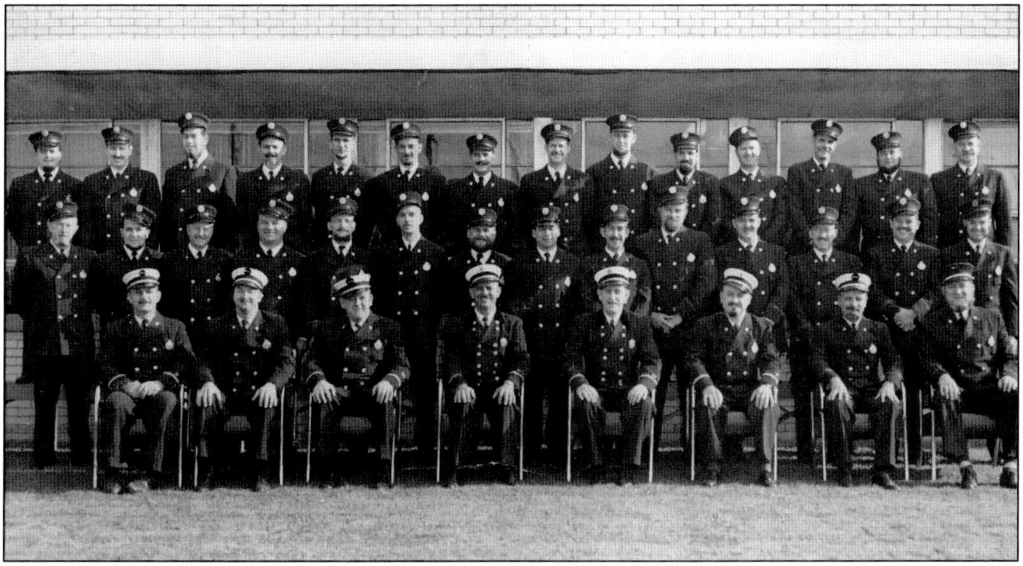

FIREMEN IN GOLDEN JUBILEE, 1956. In 1956, the city of Gary observed its Golden Jubilee and the whole city took part. Historical items from the past were put on display in public buildings. Booklets and pamphlets were printed. Decks of cards were made bearing historical photographs instead of kings and queens. People dressed in clothing similar to that at the turn of the century, and old cars from the period were restored. Here, firemen from the downtown station display their beards and mustaches for the camera. (Courtesy of CRA.)

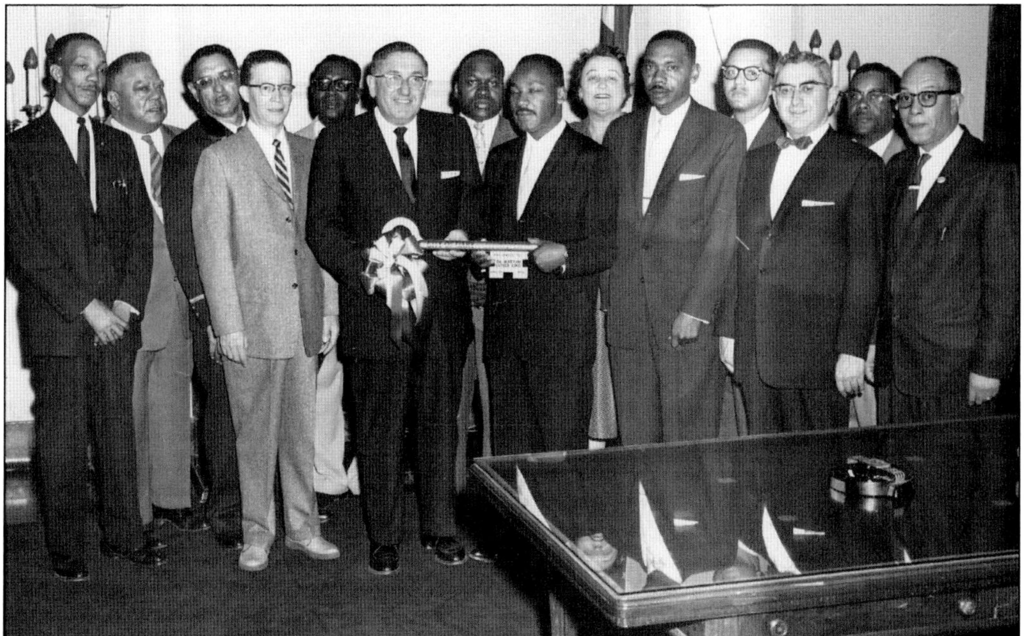

CHACHARIS WITH DR. KING, 1960. In the early 1960s, Dr. Martin Luther King Jr. came to Gary to discuss the city's progress in open housing and civil rights laws. He had meetings with the mayor and city council and was later presented the key to the city. He is pictured here meeting with Mayor George Chacharis and the city administration along with members of the National Association for the Advancement of Colored People. (Courtesy of CRA.)

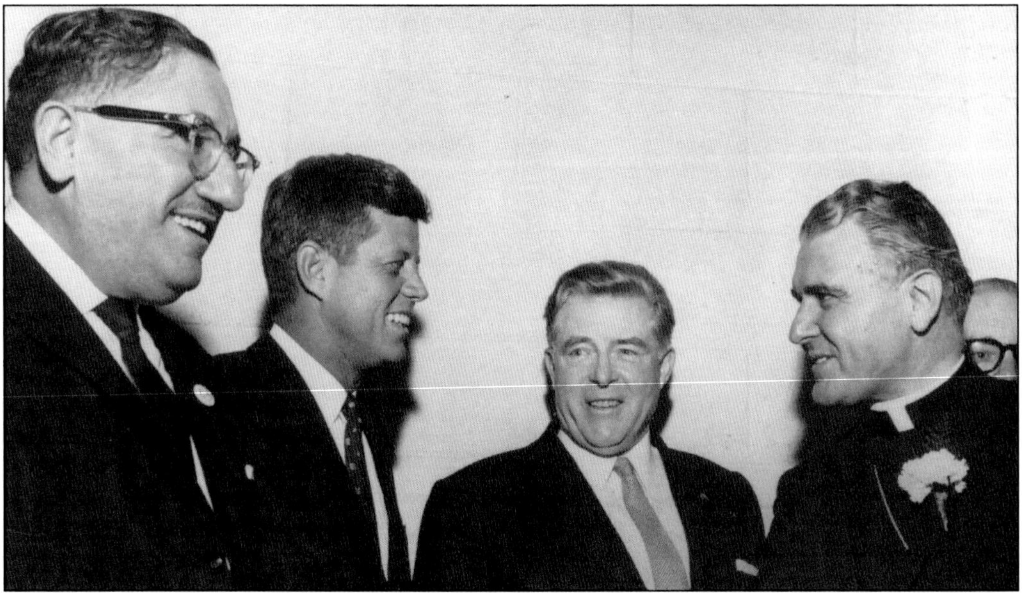

JFK IN GARY, 1960. In the 1960 presidential election, John F. Kennedy defeated fellow Democrats Lyndon Johnson, Hubert Humphrey, and Adlai Stevenson to win the party's nomination. Like his Republican opponent, Richard M. Nixon, Kennedy traveled across the country to campaign. Gary was a stop with its large blue-collar, pro-union, and Catholic population. From left to right are Mayor George Chacharis, Kennedy, 1st District congressman Ray J. Madden, and Bishop Andrew Grutka at the Diocese of Gary. (Courtesy of CRA.)

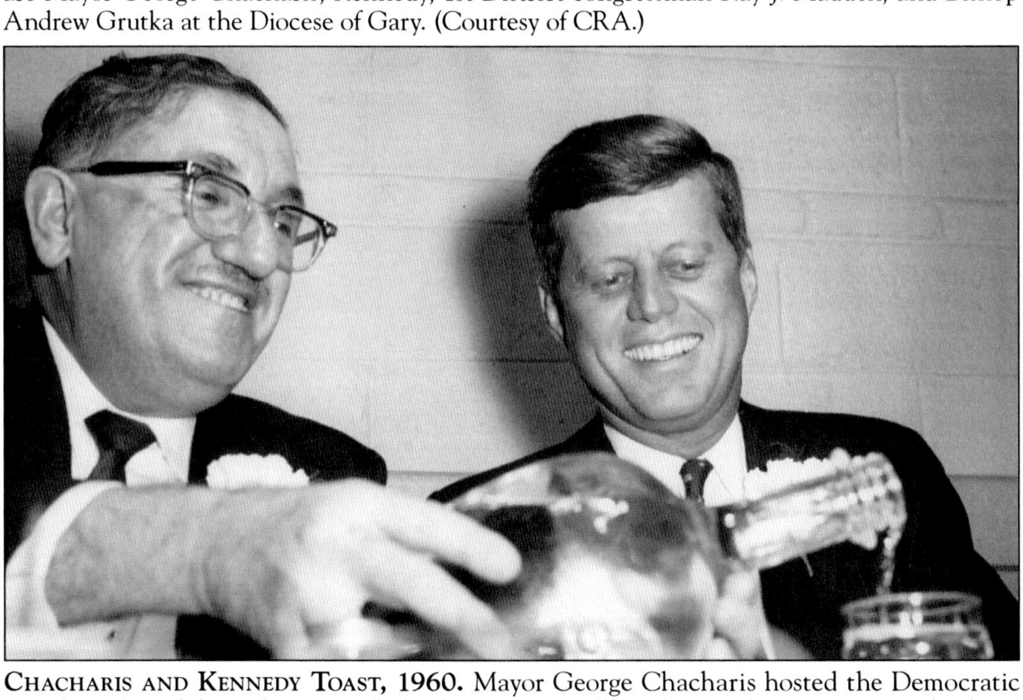

CHACHARIS AND KENNEDY TOAST, 1960. Mayor George Chacharis hosted the Democratic presidential candidate John Kennedy for dinner and a rally at the Hotel Gary in 1960. Chacharis went all out to help the young senator win over the voters of Northwest Indiana as the GOP was still strong in areas throughout the region in the 1960s. Here, the mayor is shown pouring a drink for a toast for a democratic victory. One wonders if it was ouzo or Irish whiskey. (Courtesy of CRA.)

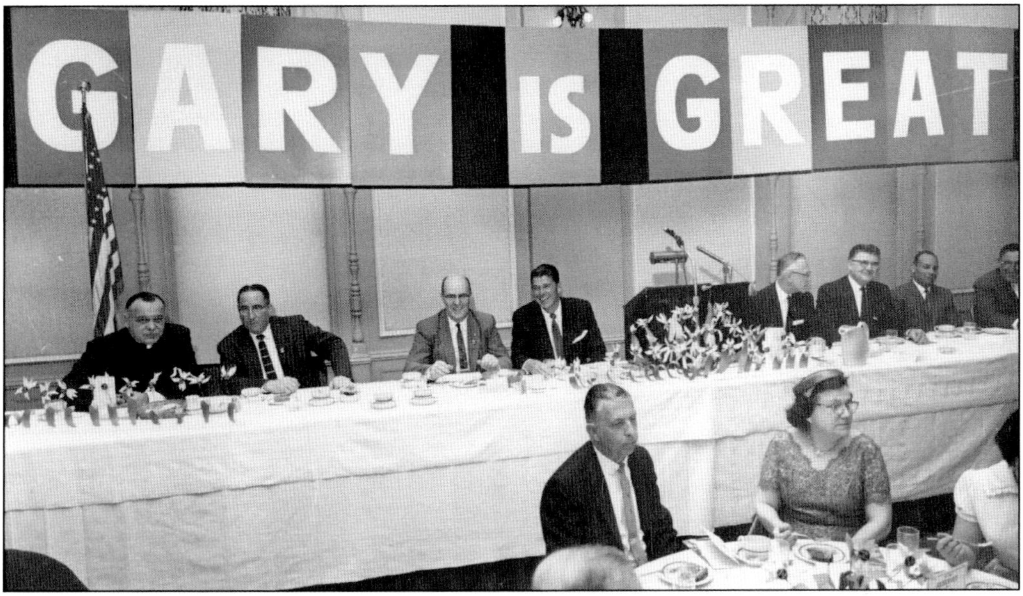

REAGAN IN GARY, 1960. While the Democratic Party's nominee for president, John F. Kennedy, campaigned in Gary and the Midwest, the Republicans were not throwing in the towel on Lake County. A number of well-known Republicans were calling in the help of Vice Pres. Richard M. Nixon. Hollywood actor and former Democrat Ronald Reagan came to rally local Republicans for Nixon at a dinner in the Hotel Gary. Reagan later served as the governor of California, and in 1981 was inaugurated as the nation's 40th president. (Courtesy of CRA.)

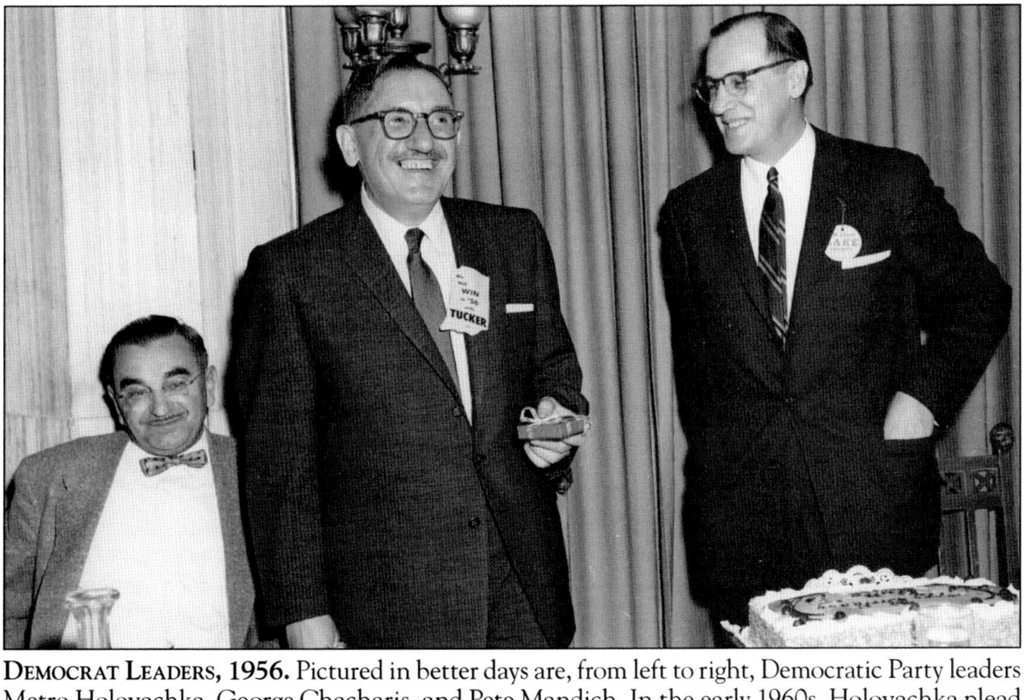

DEMOCRAT LEADERS, 1956. Pictured in better days are, from left to right, Democratic Party leaders Metro Holovachka, George Chacharis, and Pete Mandich. In the early 1960s, Holovachka plead guilty to tax evasion and spent two years in federal prison. Chacharis stepped down as mayor and plead guilty to tax evasion charges. (Courtesy of CRA.)

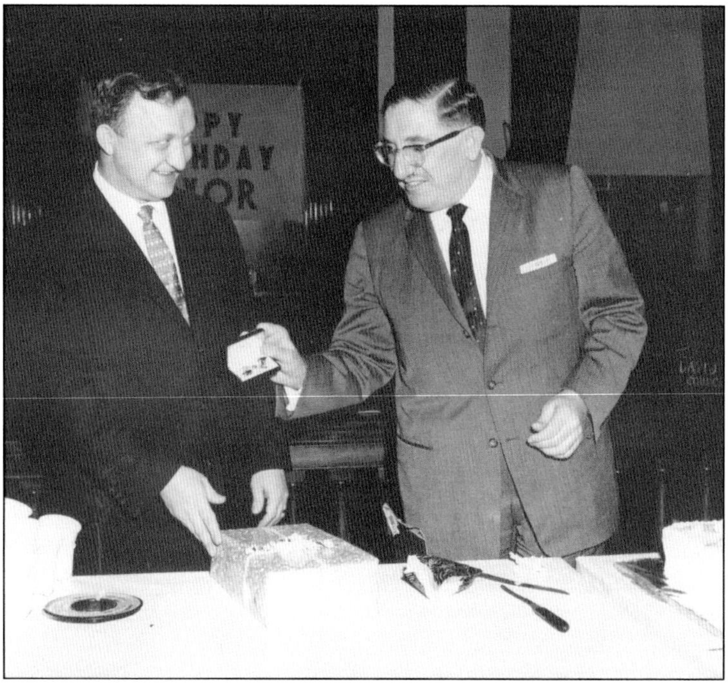

CHACHARIS AND JOHN VISCLOSKY, 1962. John Visclosky is pictured with Mayor Chacharis at his birthday party in 1962. Visclosky, former ironworker and World War II veteran, served as deputy city controller in the Chacharis administration until the indictment and conviction of the mayor. Visclosky then served the remainder of Chacharis's term and was succeeded by Judge A. Martin Katz in 1963. (Courtesy of CRA.)

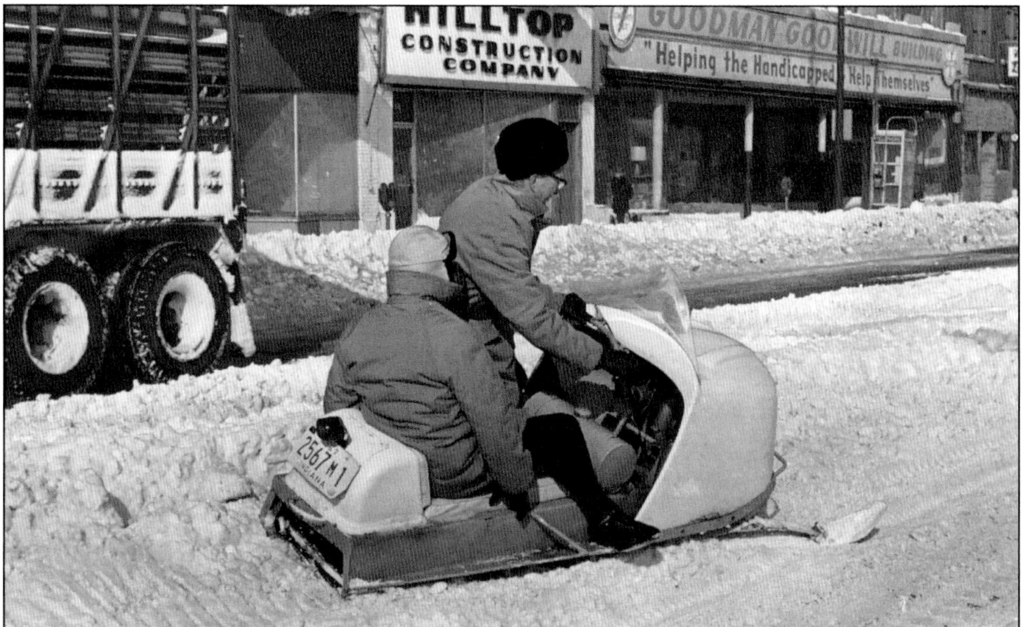

MAYOR KATZ, BLIZZARD OF 1967. During the great blizzard of January 1967, travel around Gary and the Calumet Region was nearly impossible. The storm, which began in the morning of January 26, dumped more than two feet of snow in 24 hours. Schools let out early that Thursday and would not reopen for two weeks. A cold spell followed the big snow drop, but more snow fell during the next week. The National Guard came to Gary the following week as snow was chest-high in most areas. To get around, Mayor A. Martin Katz traveled the city by snowmobile to direct snow removal work. He is shown in the 700 block of Broadway. (Courtesy of CRA.)

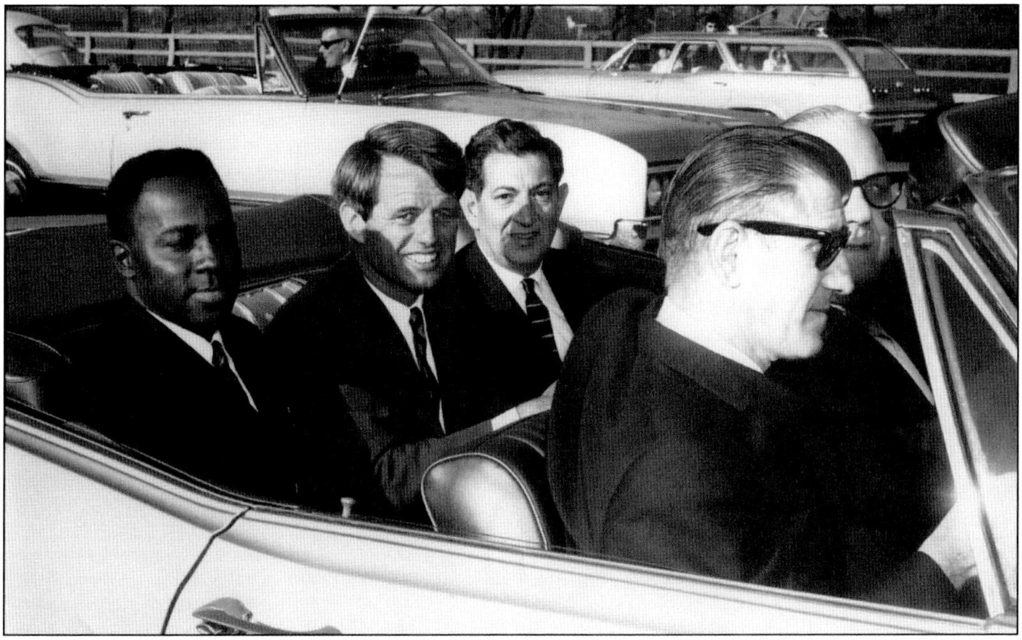

Bobby Kennedy in Gary, 1968. In 1968, Lyndon Johnson decided not to run for another term as president. The war in Vietnam made him realize that he probably would not get his party's nomination. Robert Kennedy threw his hat into the ring, facing four other Democrats in the primaries. Kennedy made two campaign stops in Gary, the first at a packed house in Memorial Auditorium. He is pictured with Richard Gordon Hatcher, who became the city's first African American mayor in the election of 1967. (Courtesy of CRA.)

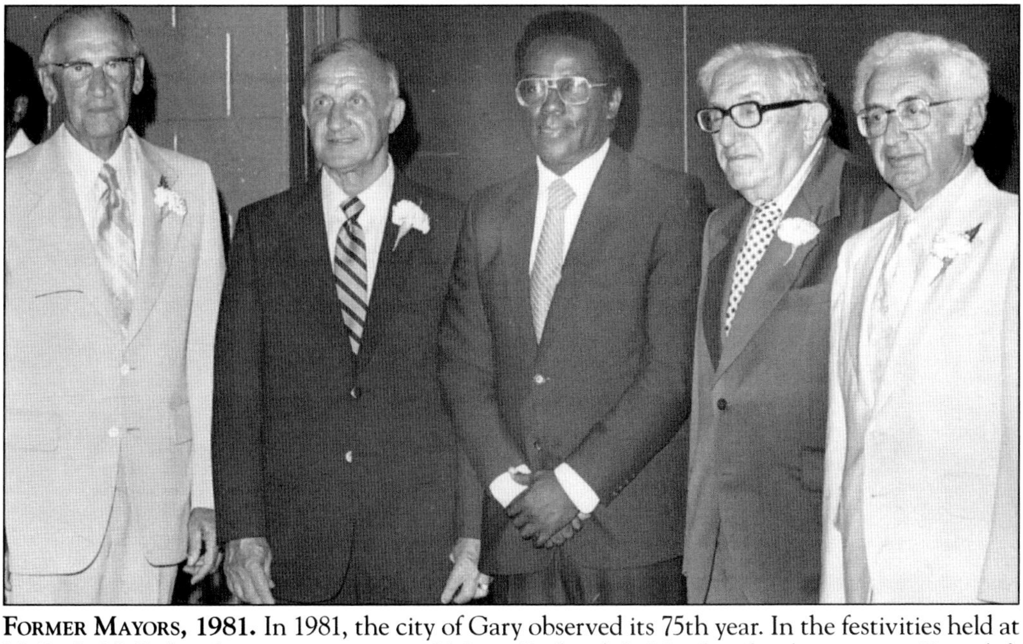

Former Mayors, 1981. In 1981, the city of Gary observed its 75th year. In the festivities held at McBride Hall, four former mayors joined Mayor Richard G. Hatcher for the reunion. From left to right are Peter Mandich, John Visclosky, Hatcher, George Chacharis, and A. Martin Katz. (Courtesy of CRA.)

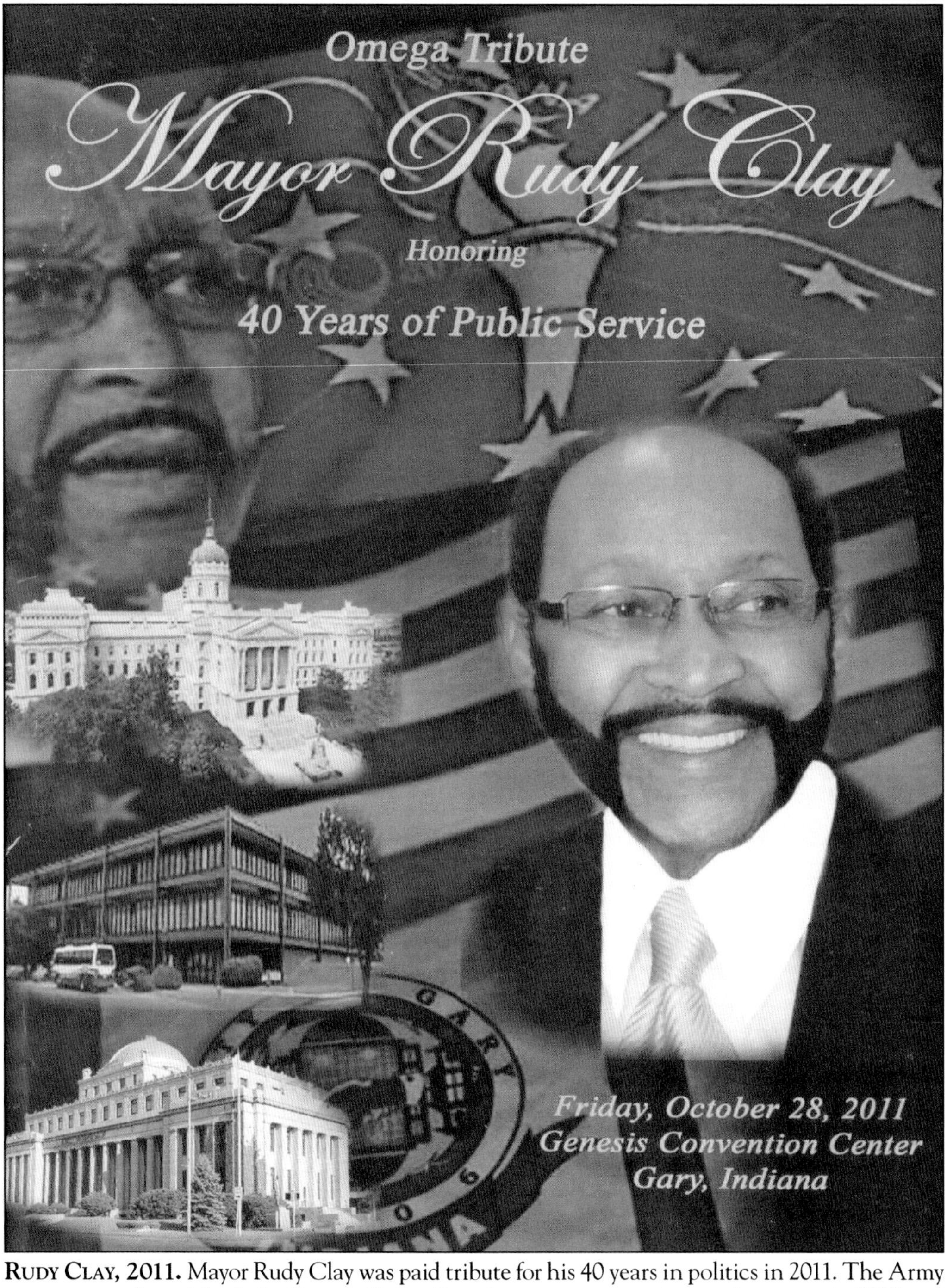

Rudy Clay, 2011. Mayor Rudy Clay was paid tribute for his 40 years in politics in 2011. The Army veteran, member of the Lake County Council, and state representative ran for mayor in 2007 and won. Because of health issues, he did not run for reelection in 2011. This image was part of the dinner program given in his honor at the Genesis Center in Downtown Gary. (Courtesy of CRA.)

Four

Downtown Businesses

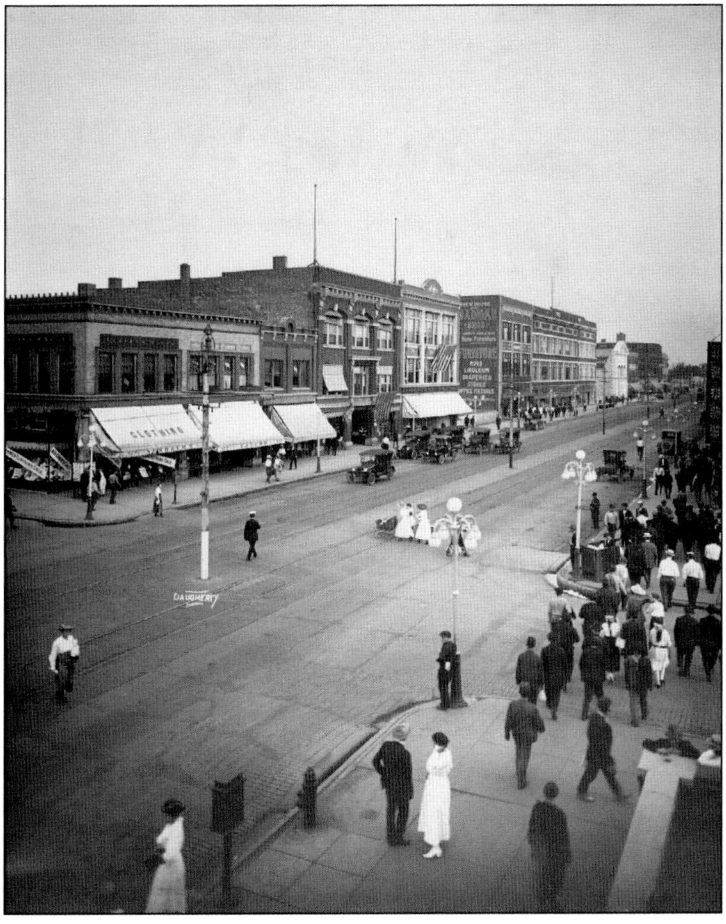

Broadway, 1916. Downtown Gary was a far different place in 1916. Life was much simpler and more laid back. Besides, the war in Europe was an ocean away and not America's conflict. Gary's main street, Broadway, had gone through some major business growth between Fifth Avenue and the Wabash Railroad tracks, but many lots were still vacant. Pictured is Sixth Avenue, looking southeast. The building that would hold Goldblatt's was owned by another business. To its left, Radigan Bros. Furniture was in business. There was little traffic on the street, and no traffic signals. Instead, a traffic cop directed the few automobiles. (Courtesy of CRA.)

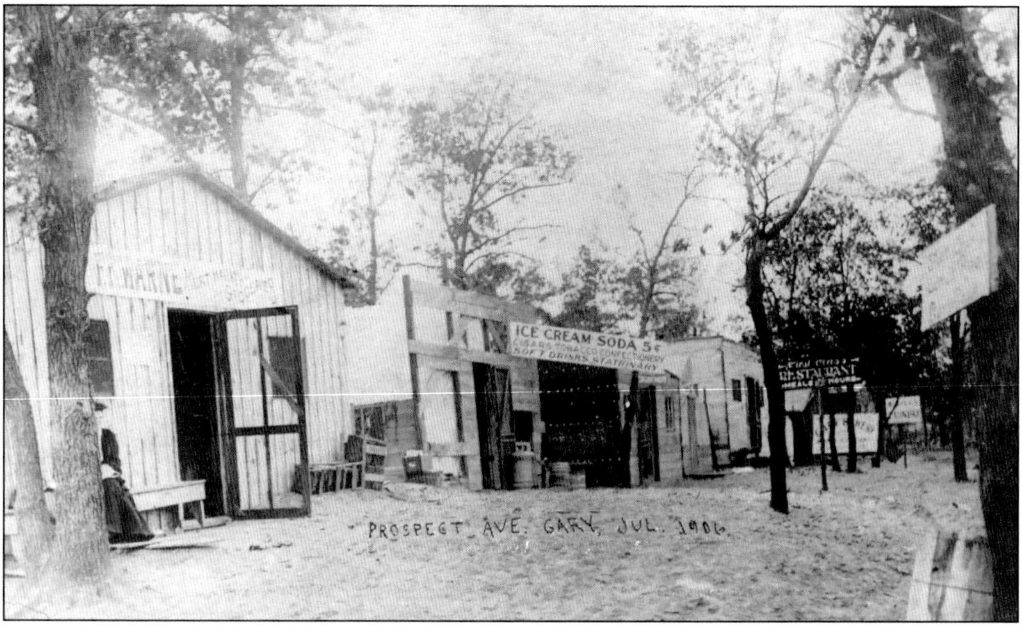

EARLY BUSINESS, 1906. Before the city of Gary was surveyed and streets were laid out, small businesses emerged outside the US Steel properties. The early stores at first provided workers building the plant with simple necessities. Work pants, shoes, and socks were in demand. Provisions such as food, blankets, and even alcohol were available. In 1906, Gary was more like a Western frontier town in the gold or silver areas of Colorado or California. (Courtesy of CRA.)

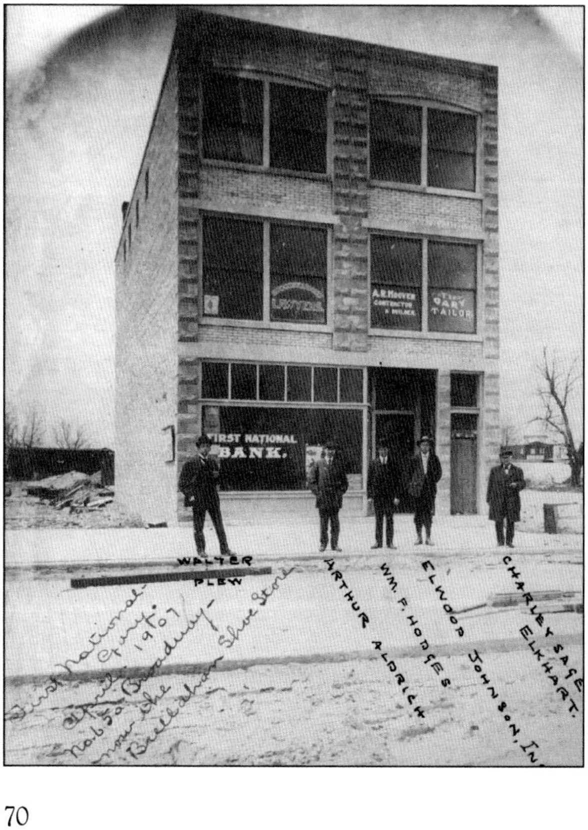

FIRST NATIONAL BANK, 1907. One of the early banks that opened in Gary soon after the mill began operations was the First National Bank, located near Fifth Avenue and Broadway. Before the bank opened, the only vaults available to store cash were found at the main office building at US Steel. At the time, men were paid in cash. Checks were used a few years later. The men, for the most part, cashed those checks at their local saloon. Cashing checks at the bar was common in steel and mining towns. The men trusted the local saloon owner more than the bank. (Courtesy of CRA.)

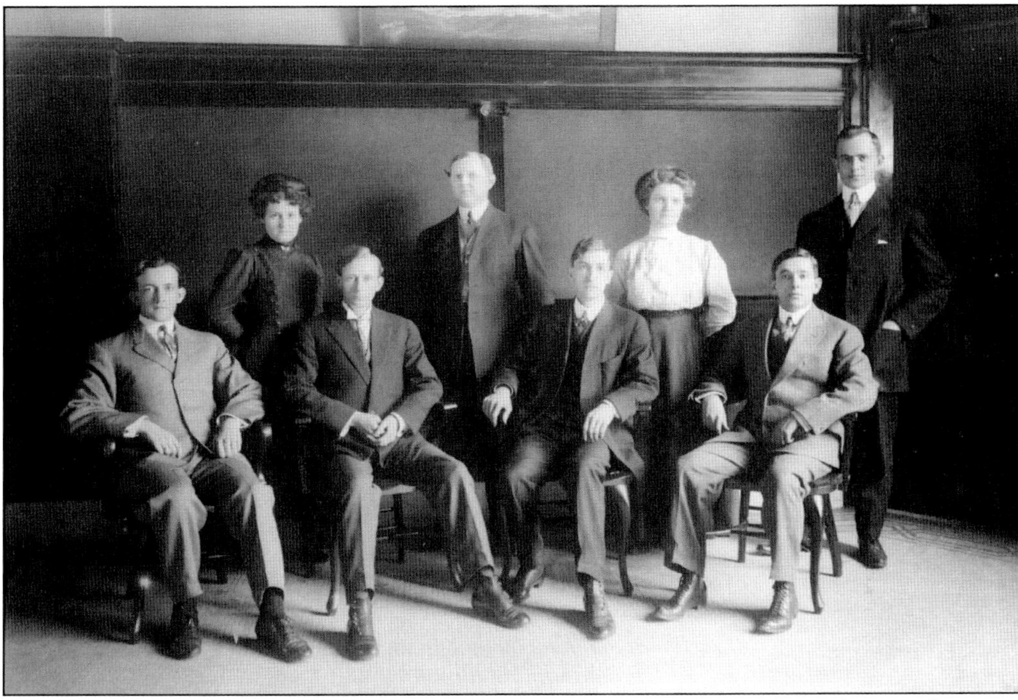

First National Bank Staff. Pictured is the staff of the First National Bank around 1910. In those days, few if any women worked as bank clerks, secretaries, or managers. (Courtesy of CRA.)

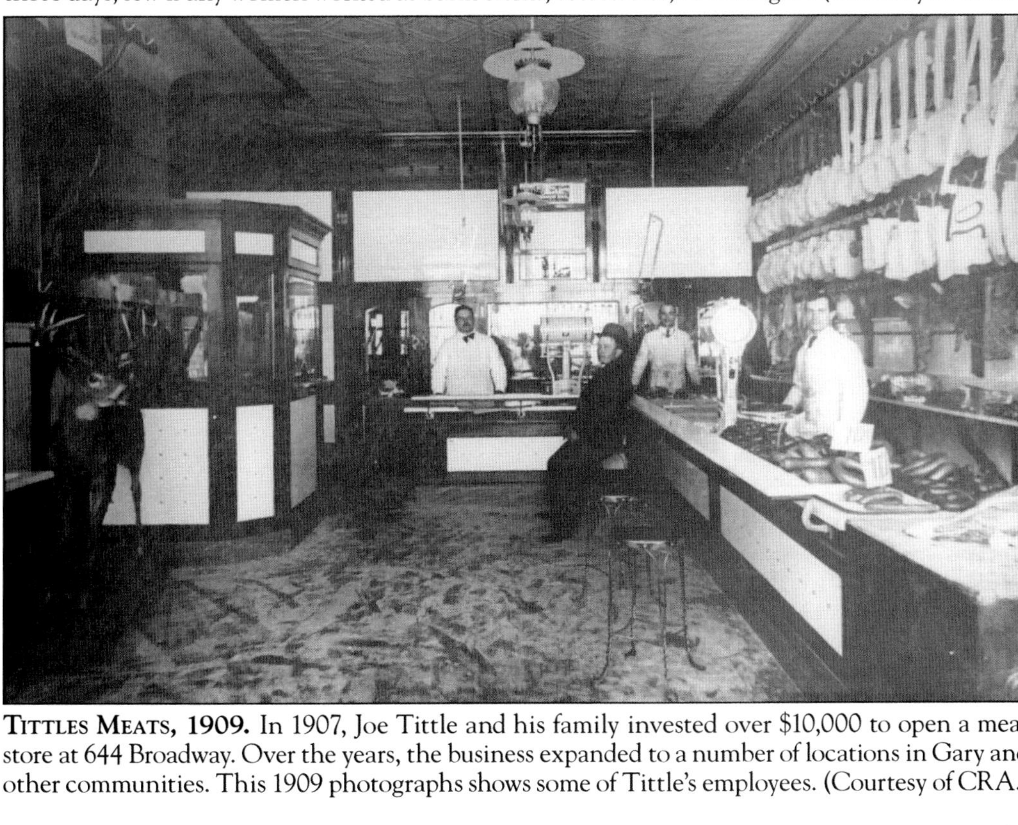

Tittles Meats, 1909. In 1907, Joe Tittle and his family invested over $10,000 to open a meat store at 644 Broadway. Over the years, the business expanded to a number of locations in Gary and other communities. This 1909 photographs shows some of Tittle's employees. (Courtesy of CRA.)

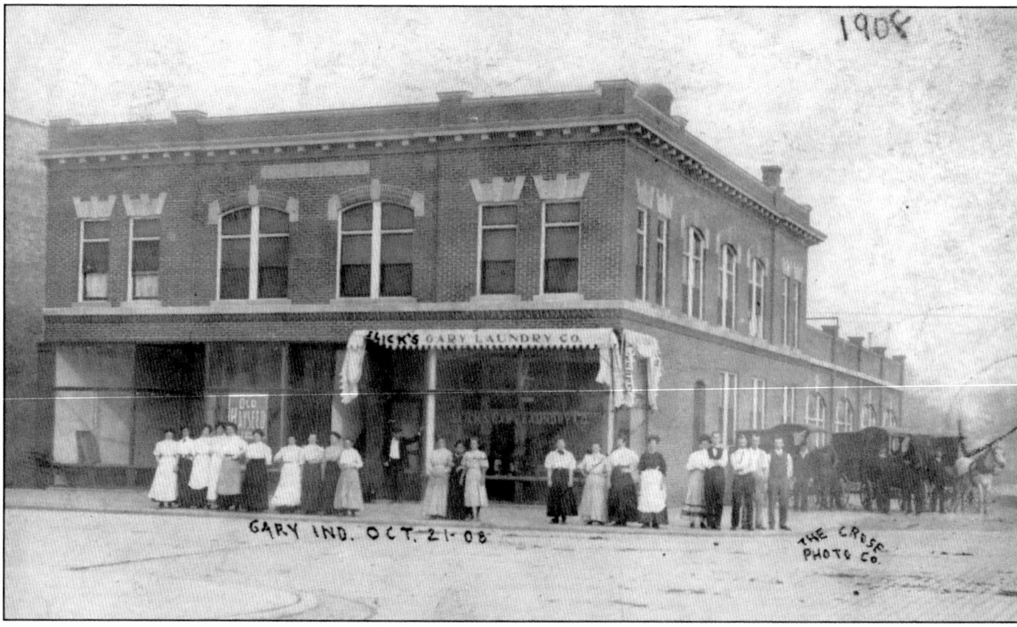

SLICK'S GARY LAUNDRY CO., 1908. At the northwest corner of Fifth Avenue and Massachusetts Street was Slick's Gary Laundry Co. The downtown operation occupied much of that corner. The business handled most of the city's cleaning, pressing, and alterations. Slick's had branch operations in the Miller, Glen Park, Tolleston, and Midtown neighborhoods until the 1980s. (Courtesy of CRA.)

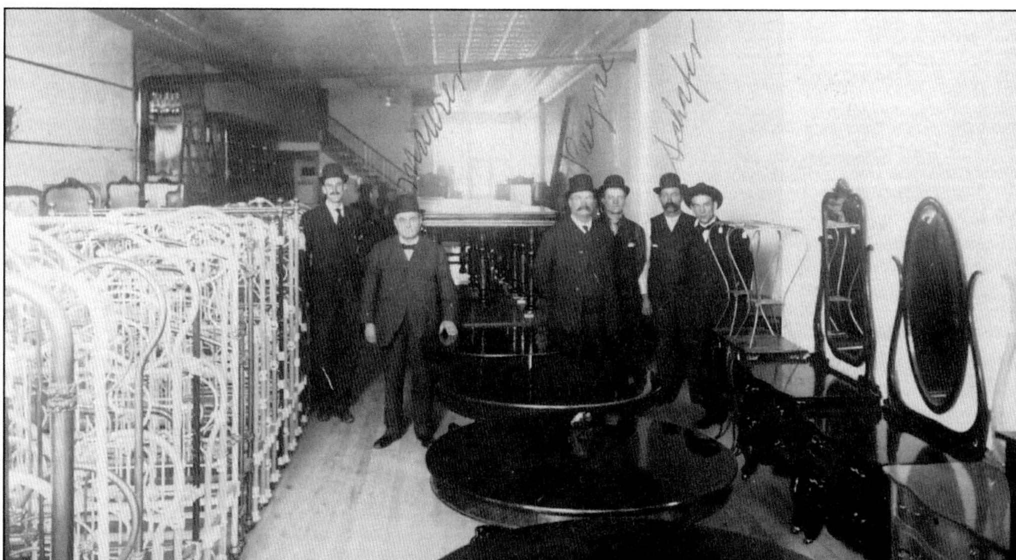

PAINE HARDWARE, 1912. In the days before Menards and Home Depot, the neighborhood hardware store was there to help folks doing their home repairs. Paine Hardware, located downtown, provided each customer with the personal assistance they needed. An attendant would personally look for the needed item, and if not in the store, they would order it or refer the customer to another store in the area. Here, the staff poses with bedroom furniture, mattress springs, and dressers. Hats were common in the years before World War I. (Courtesy of CRA.)

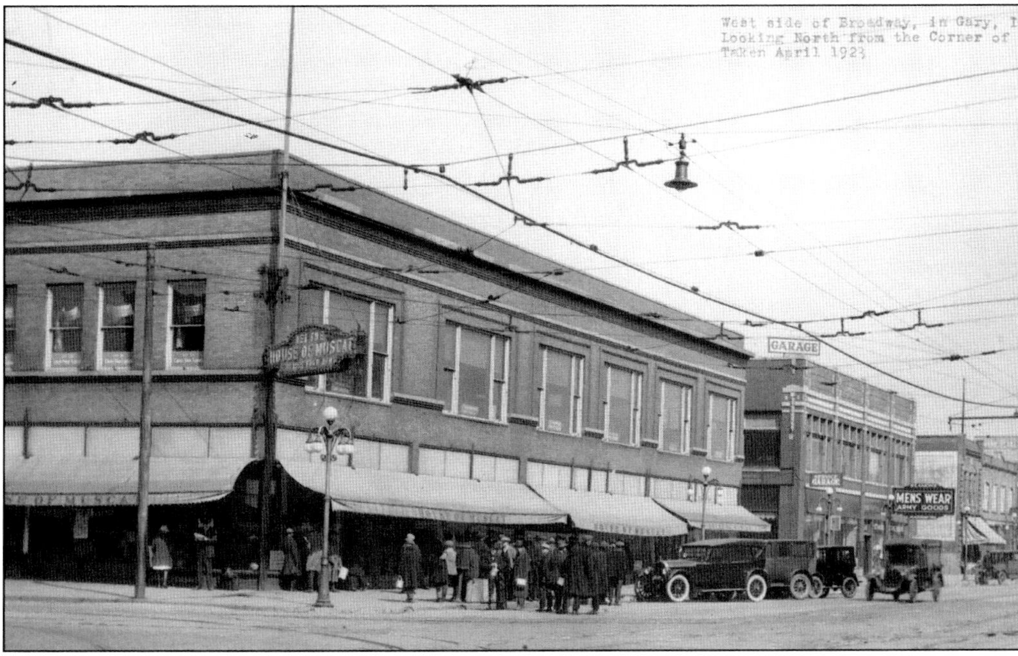

Fifth Avenue and Broadway, West Side, 1923. During the 1920s, Gary's downtown area and parts of West Fifth Avenue saw a great deal of growth and change. Along the west side of the 400 block of Broadway, the men's store House of Muscat occupied the northwest corner of that intersection. Later, the building became Henry C. Lytton & Sons. It was torn down in the 1970s. (Courtesy of CRA.)

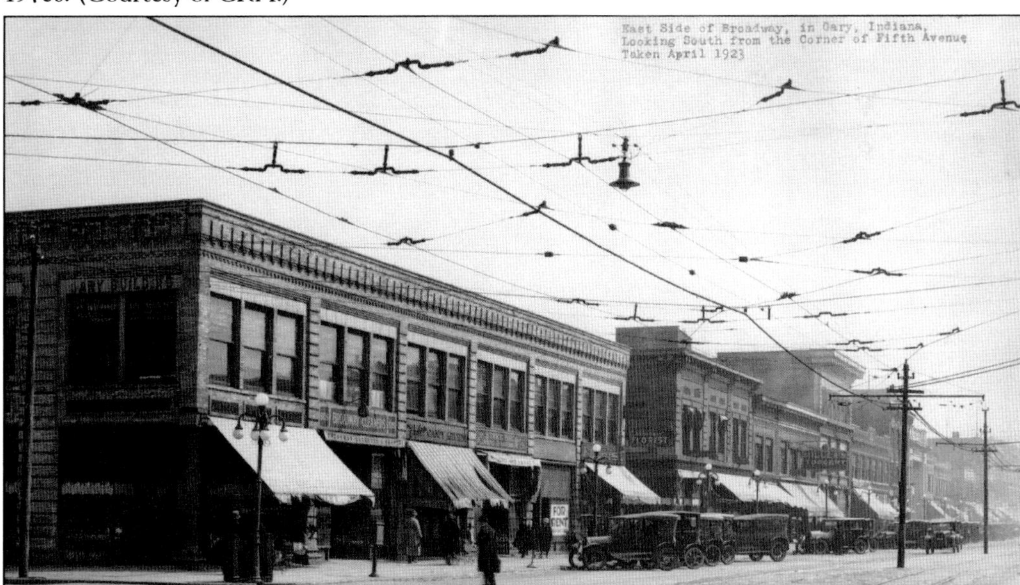

East Side of Fifth Street and Broadway, 1923. Looking south from the intersection of Fifth Avenue and Broadway, the Gary Building, as it was named, is pictured. It occupied the entire corner. A number of businesses were located there including cleaners, a bakery, cigar store, and shoe repair business. Just south were the Miller's Toggery and a fur store. This photograph was taken in April 1923. (Courtesy of CRA.)

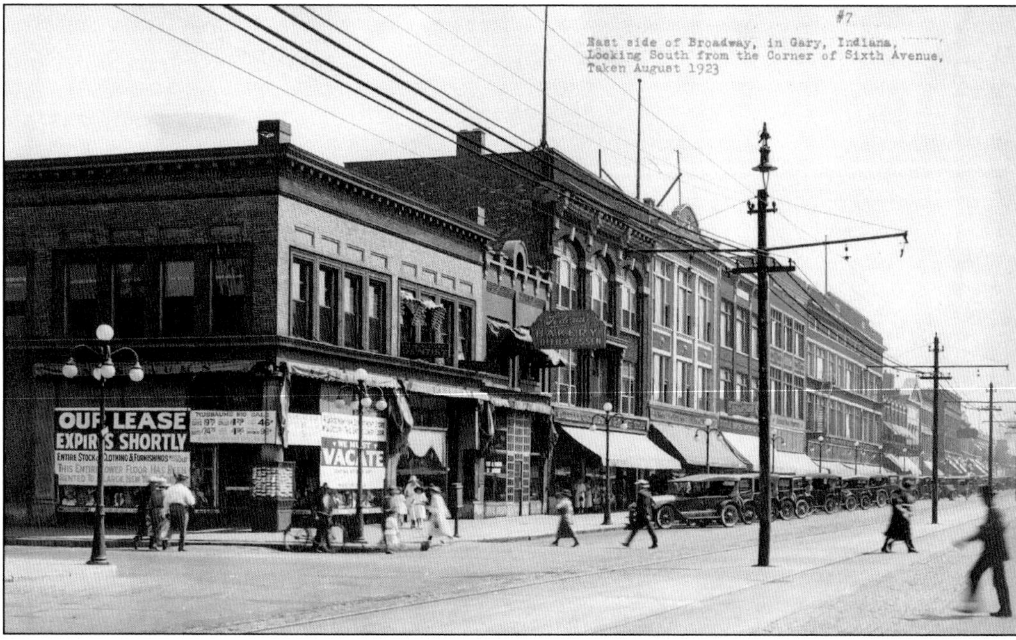

East Side of Broadway 600 Block, 1923. Along the east side of Broadway in the 600 block were department stores, clothing stores, and furniture businesses. Various offices occupied the upper floors. In this photograph, taken in late summer 1923, the seventh building was Kolbacks. Later, the Goldblatt brothers purchased it. Goldblatt Bros. became a popular department store where one always made a good deal. (Courtesy of CRA.)

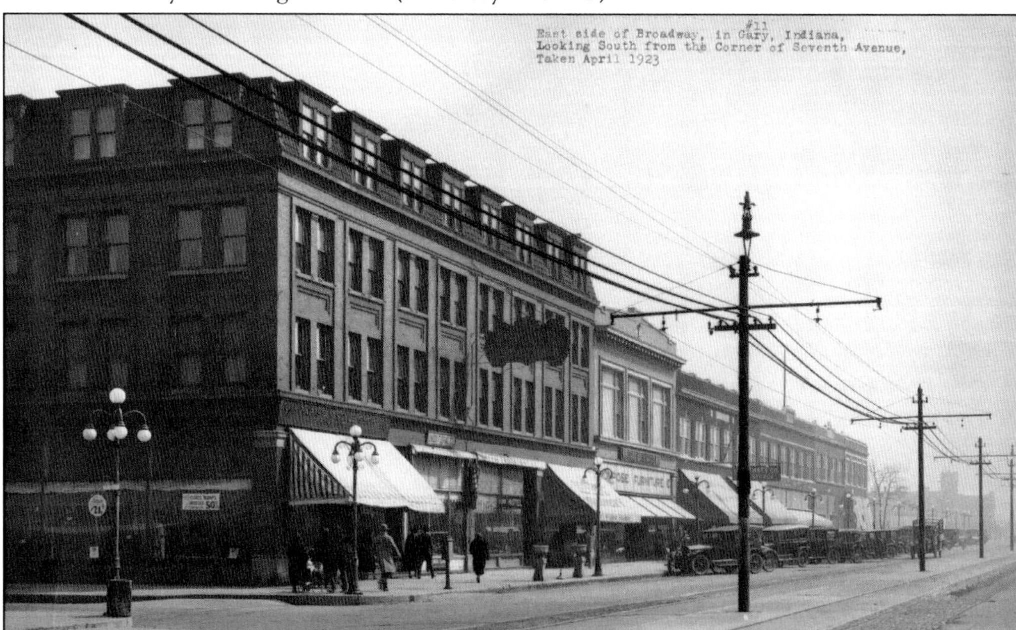

Victoria Hotel, 1923. In the 1920s, the Victoria Hotel stood on the southeast side at Seventh Avenue and Broadway. Two doors down was the J.C. Penney store, which moved to the west side of the street. The Victoria Hotel was torn down in the early 1950s. W.T. Grant purchased the site and built a discount store that went out of business in the 1970s. (Courtesy of CRA.)

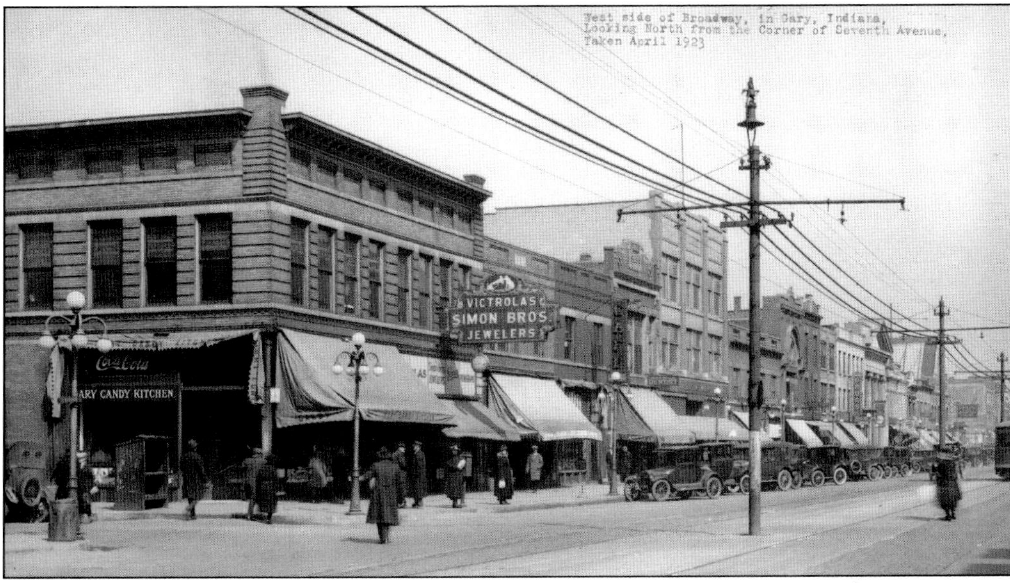

NORTHWEST CORNER OF SEVENTH AVENUE AND BROADWAY, 1923. In this 1923 photograph, the northwest corner of Seventh Avenue and Broadway is shown. Simon Bros. Jewelers is the corner business that sold victrolas, or record players. The building's ground level became a record shop in the early 1960s. (Courtesy of CRA.)

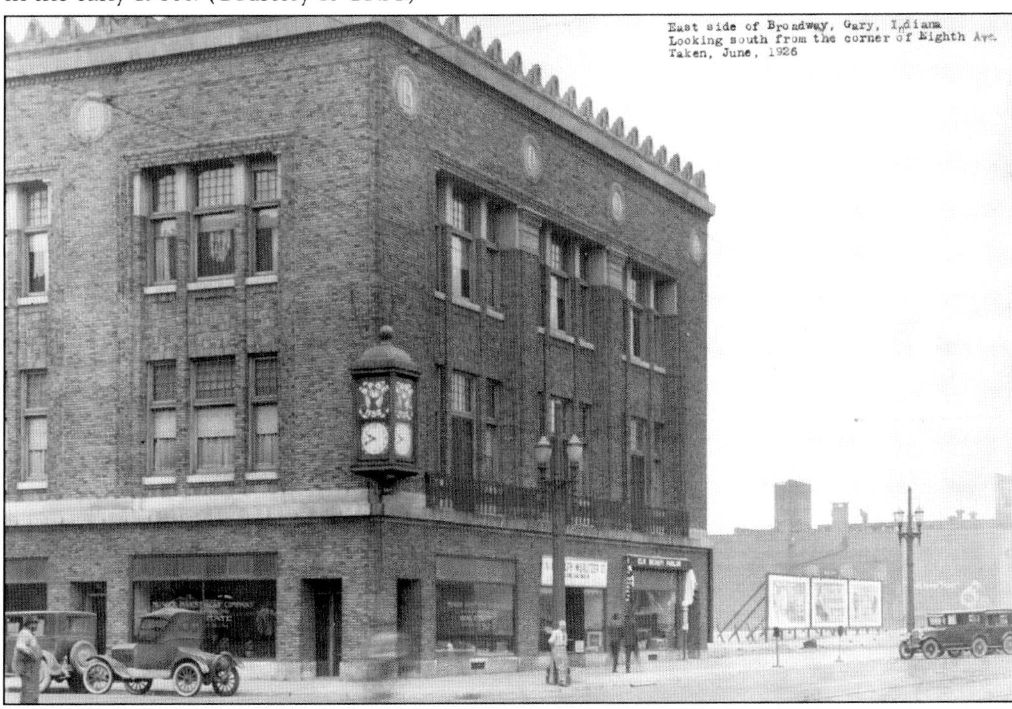

H. GORDON & SONS DEPARTMENT STORE. In the 1920s, H. Gordon & Sons purchased the three-story building at the intersection of Eighth Avenue and Broadway to expand into the Gary market. The original H. Gordon & Sons was in Whiting, Indiana. The building along with its classic clock became a landmark in the downtown area. A fourth floor was added in the 1930s. (Courtesy of CRA.)

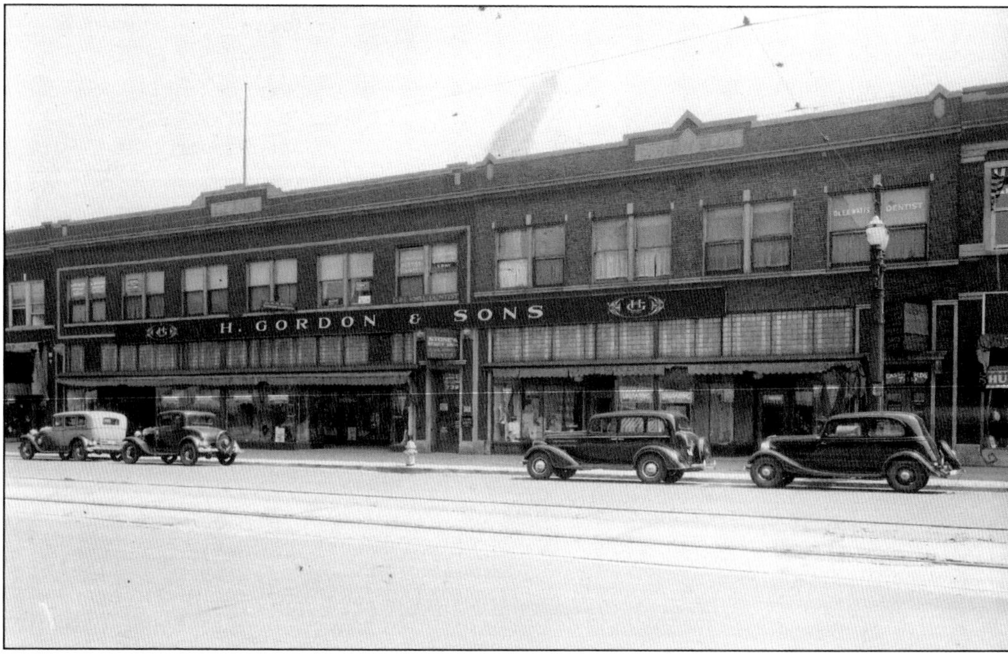

H. GORDON & SONS STORE, WHITING. H. Gordon & Sons' original store was located along 119th Street in Whiting, Indiana. It is pictured here in the 1930s. (Courtesy of CRA.)

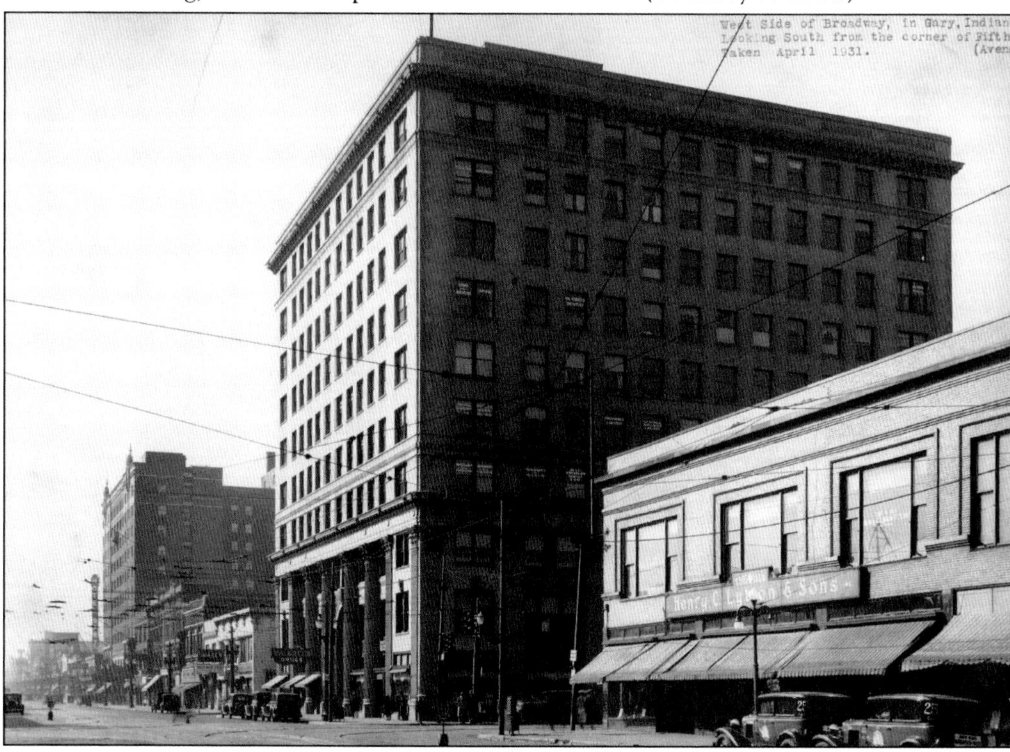

GARY STATE BANK, 1925. The Gary State Bank, later the Gary National Bank, is shown here in 1925. It was located at the southwest corner of Fifth Avenue and Broadway. During the Great Depression, it was the only financial institution in the city to keep its doors open. (Courtesy of CRA.)

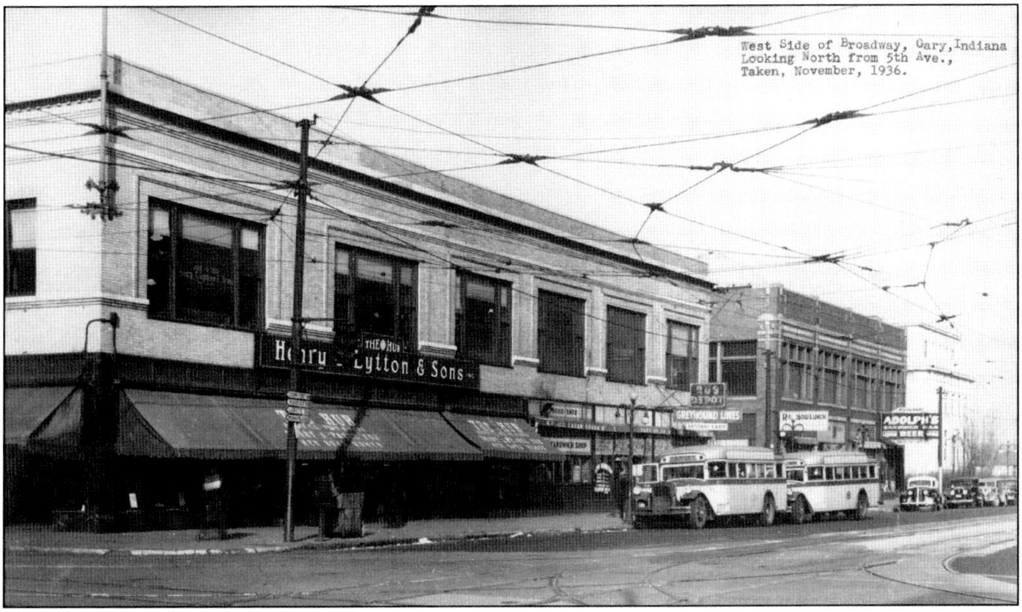

HENRY C. LYTTON & SONS, 1936. In the 1930s, the House of Muscat moved to another location downtown while the Henry C. Lytton & Sons men's store purchased the Fifth Avenue building. The store sold quality men's clothing for nearly 40 years until it closed its Gary store in the late 1960s. (Courtesy of CRA.)

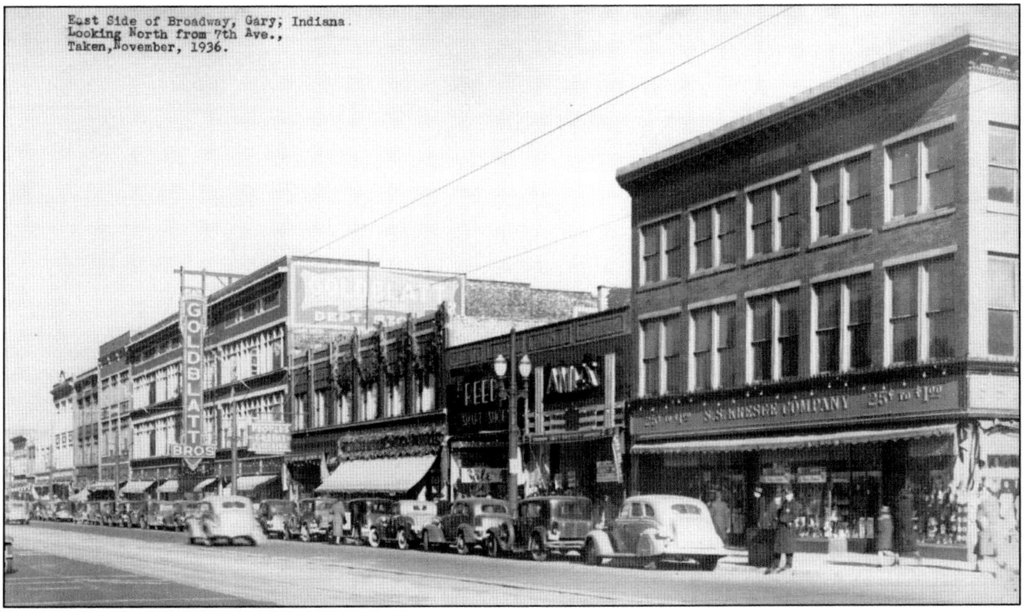

GOLDBLATT BROS., 1936. Gary's downtown merchants went all out in decorating their stores for the Christmas holidays. Stores like Sears and Goldblatt Bros. set up Lionel train displays, the latest dolls were on the shelves, and both brought in their own Santas. In this 1936 photograph, S.S. Kresge Company is shown at the corner of Sixth Avenue and Broadway. Goldblatt Bros. is a few doors down. In 1914, brothers Nate and Maurice Goldblatt opened stores in Chicago and, a few years later, Indiana. Gary's store opened in the 1930s. The business closed for the last time in 1981. (Courtesy of CRA.)

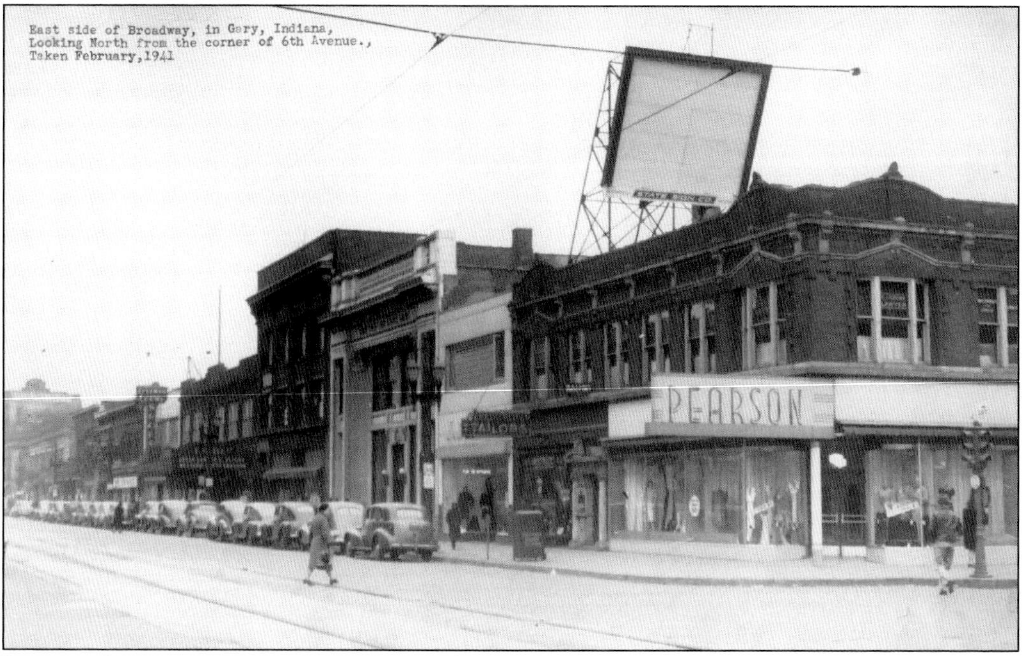

PEARSON STORE, 1941. Downtown Gary had a great selection of specialty shops. One of the popular shops for women was Pearson, located at the northwest corner of Sixth Avenue and Broadway. It offered quality ladies' goods at reasonable prices, with excellent service, for working-class families. It is pictured here in the early 1940s. (Courtesy of CRA.)

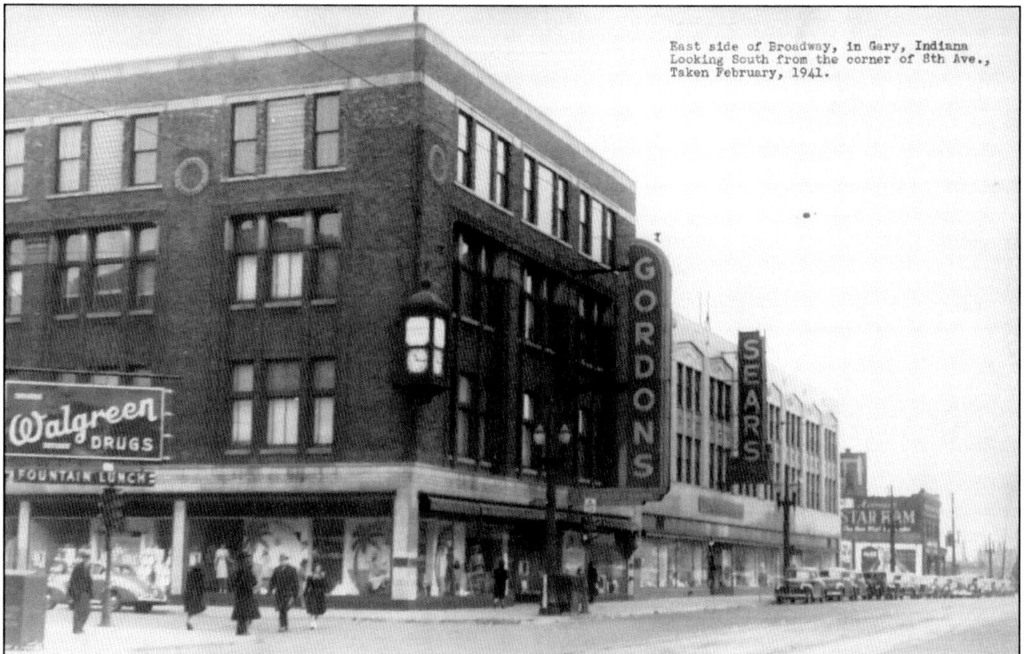

H. GORDON & SONS, 1941. Shown in February 1941 is H. Gordon & Sons with the new addition of the fourth floor. America was not yet in the war, but in another 10 months, all would change for the United States and its citizens. (Courtesy of CRA.)

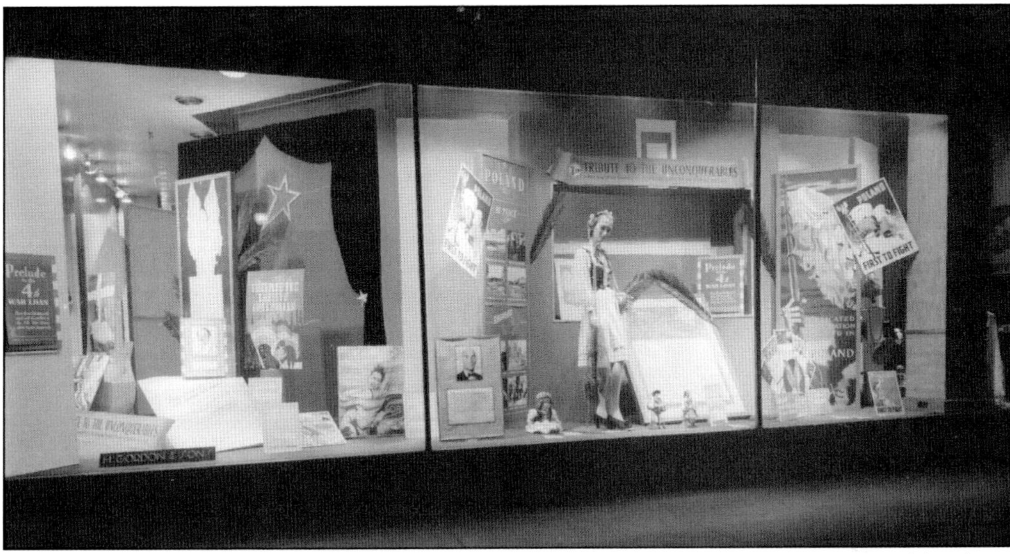

H. GORDON & SONS' DISPLAY WINDOWS, WORLD WAR II. During the war, businesses, public places, and local industries stressed a positive patriotic theme to lead the nation to victory. The windows at H. Gordon & Sons showcased mannequins dressed in the traditional clothing of those native to Poland and other states that had been conquered by Germany. The display hoped to convey support for the United States' allies in Europe and freedom for all people. (Courtesy of CRA.)

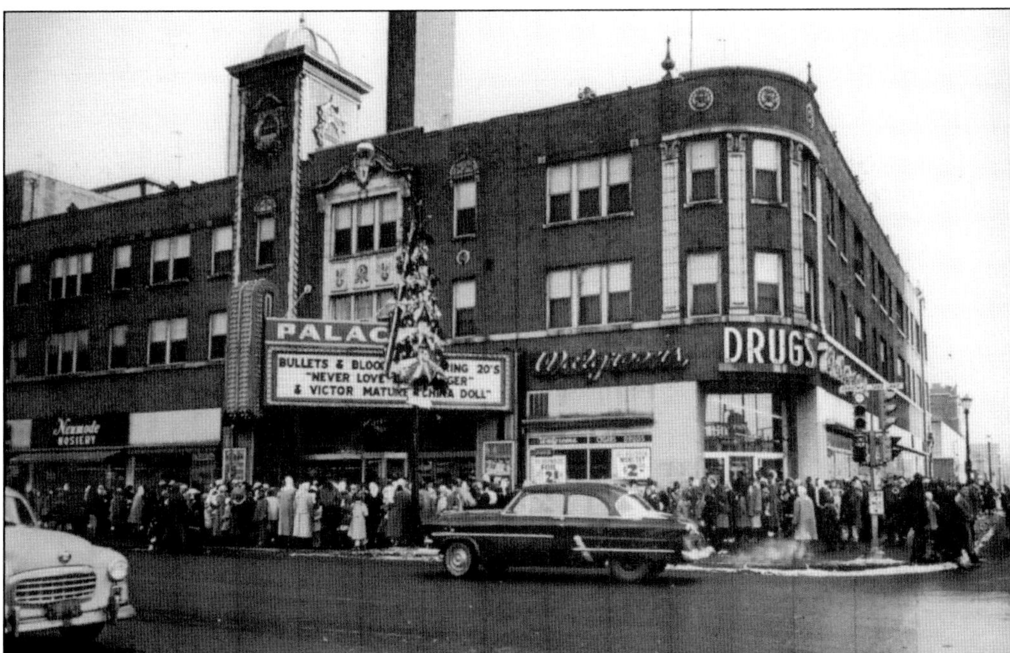

PALACE THEATER, 1950s. The Good Fellow Club, a social and charitable group sponsored by US Steel, held special events for the youngsters. In the 1950s, they hosted special bargains for tickets at the Palace Theater on Saturdays. Often, prizes were awarded for lucky ticket holders. An event was held at the Palace in December 1954. The Victor Mature movie shown on the marquee is not what the kids were going to see. (Courtesy of CRA.)

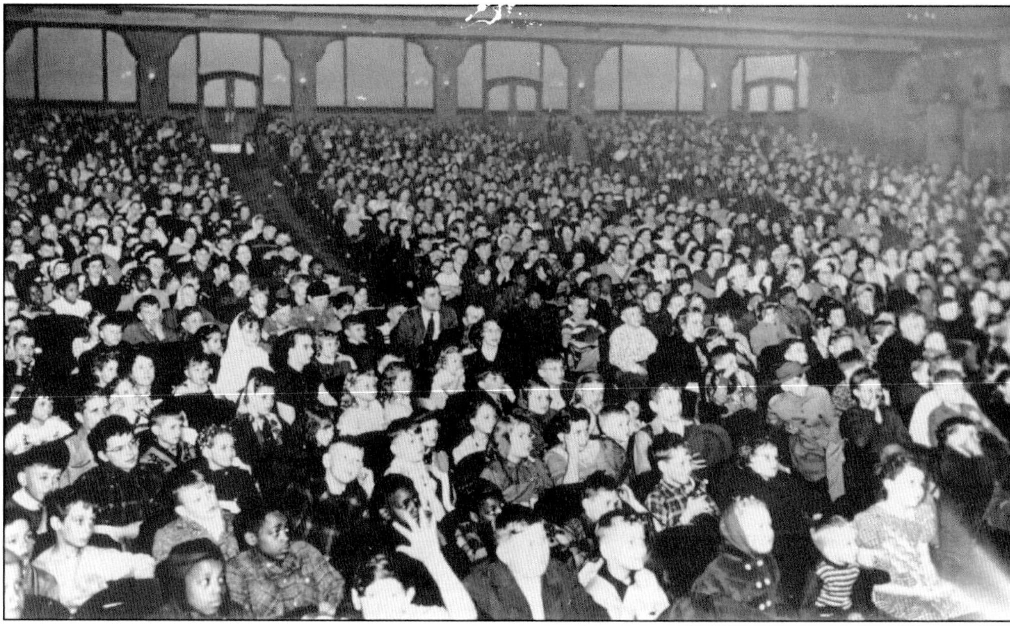

PALACE INTERIOR, 1950S. The Palace Theater was built in the splendid style of the movie houses of the 1920s. Its lobby with grand stairway and artwork was set like a European palace. The theater's seating area was like sitting in a royal courtyard with the ceiling made to look like a starry night. It sat over 1,500 people. This mid-1950s photograph shows a theater full of youngsters at the Saturday movie matinee and contest to win a new bike. (Courtesy of CRA.)

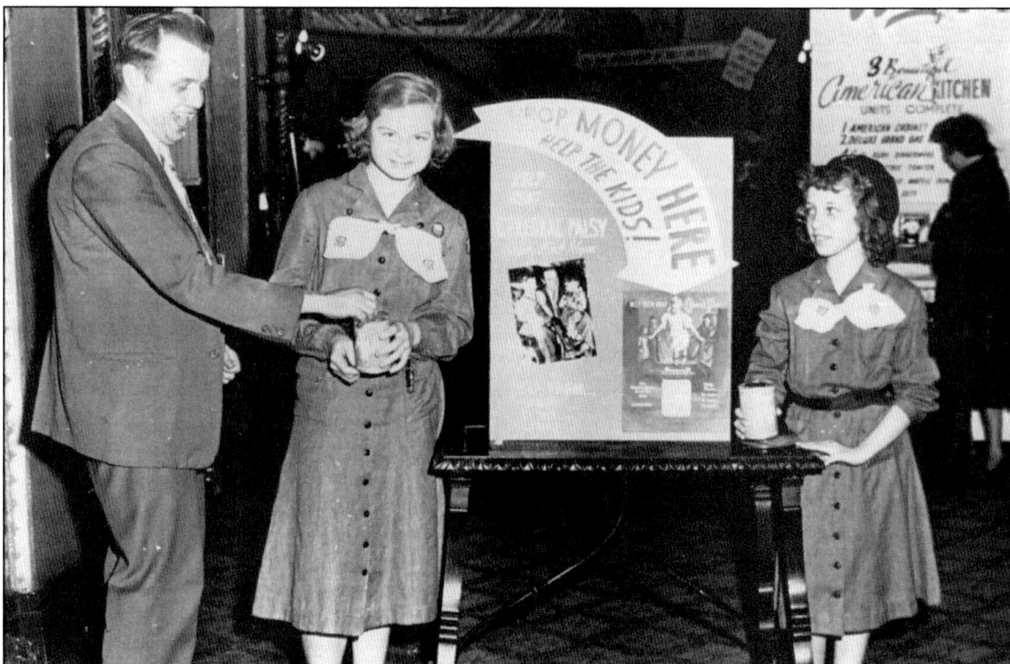

SCOUTS AT PALACE THEATER, 1950S. Like many Gary merchants, the Palace ran a number of charitable events to benefit schools, Scouting groups, and local fund drives. In the 1950s, the owners helped local Girl Scouts raise money for the fight against cerebral palsy. (Courtesy of CRA.)

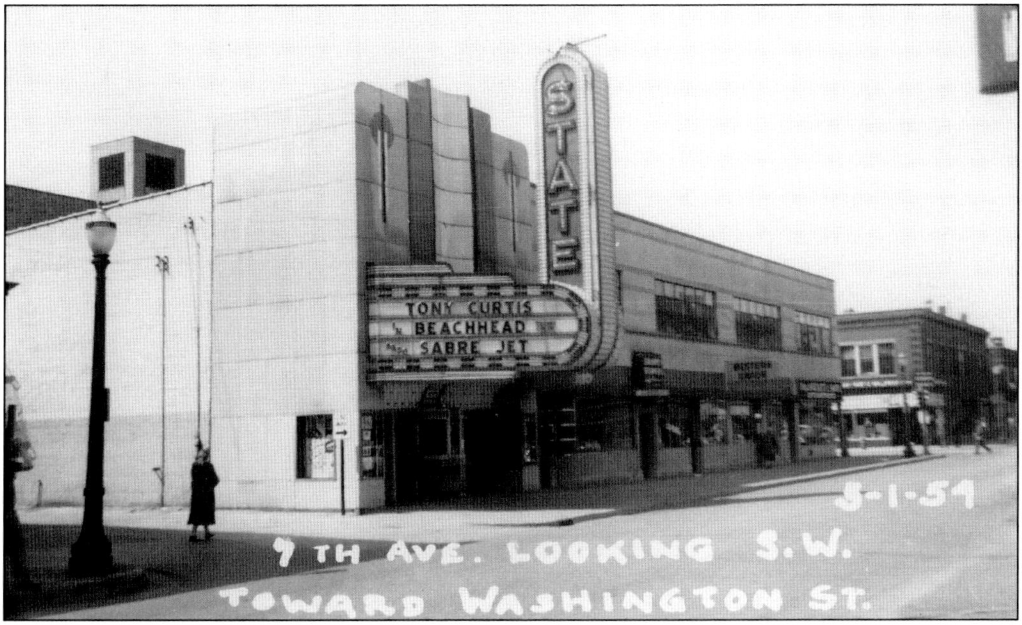

STATE THEATER, 1954 VIEW. Gary's other major downtown theater from the 1940s through the 1960s was the State, located at Seventh Avenue and Washington Street. It was built in 1937. At one time, the city had 11 movie houses. The Palace, State, Gary Theater, Fifth Avenue, and Trivoli were all in or near downtown. Others were the Indiana, Roosevelt, Ridge, and the Roxy. (Courtesy of CRA.)

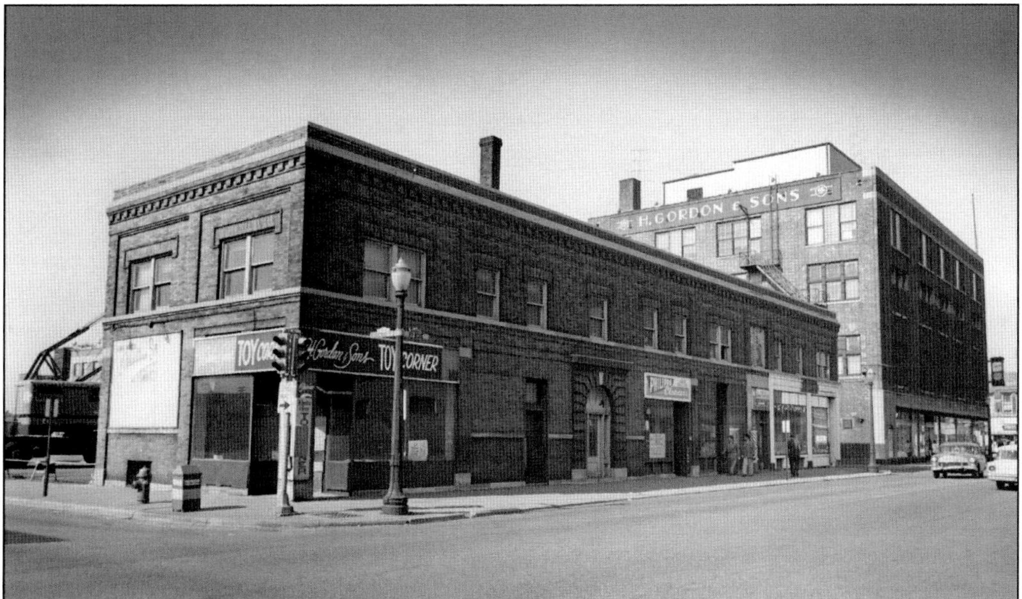

H. GORDON & SONS TOY CORNER, 1960. H. Gordon & Sons operated a toy store at the intersection of Eighth Avenue and Massachusetts Street, across from the A&P Store. The structure also held a barbershop and bakery until the building was torn down for expansion. H. Gordon & Sons built a modern two-story building that was connected to the main store by an overpass. The old building is shown in the 1950s. (Courtesy of CRA.)

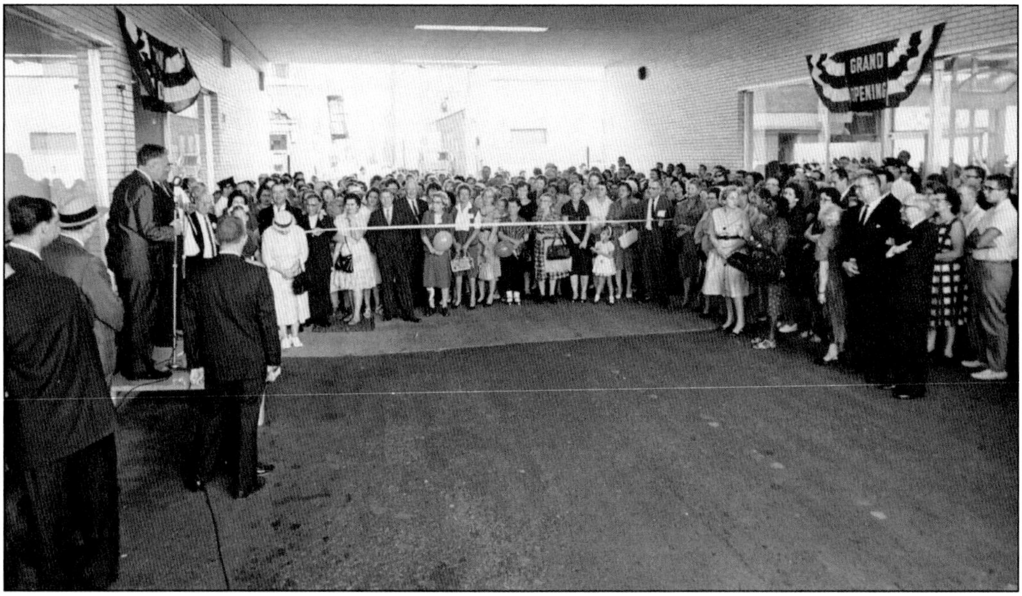

GRAND OPENING, 1960. In 1960, H. Gordon & Sons held the grand opening of its new Massachusetts Street Store by offering huge sales and promotions. Bargains were available throughout the store, not just the basement. Free items were handed out and special drawings for prizes were held. (Courtesy of CRA.)

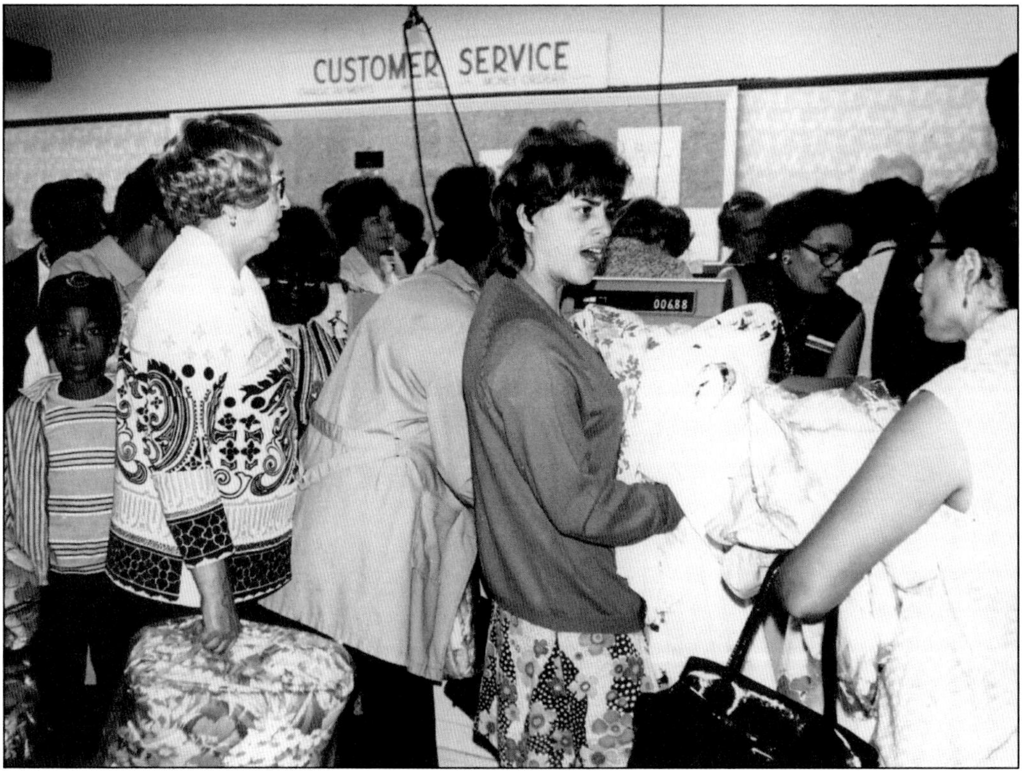

SHOPPERS, 1960S. H. Gordon & Sons drew plenty of shoppers for special sale events held at the store. A shopper is seen here trying to get the best deal. (Courtesy of CRA.)

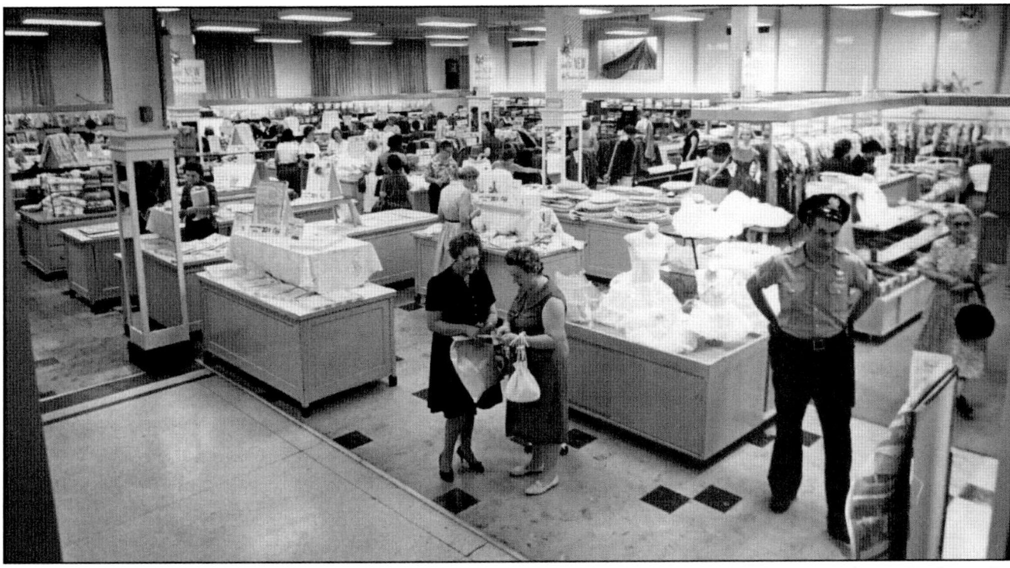

TIRED OFFICER AT H. GORDON & SONS. As the grand opening event came to an end, the sales clerks, floor managers, and staff could breathe a sigh of relief. The police officers assigned to keep order looked forward to a good rest. (Courtesy of CRA.)

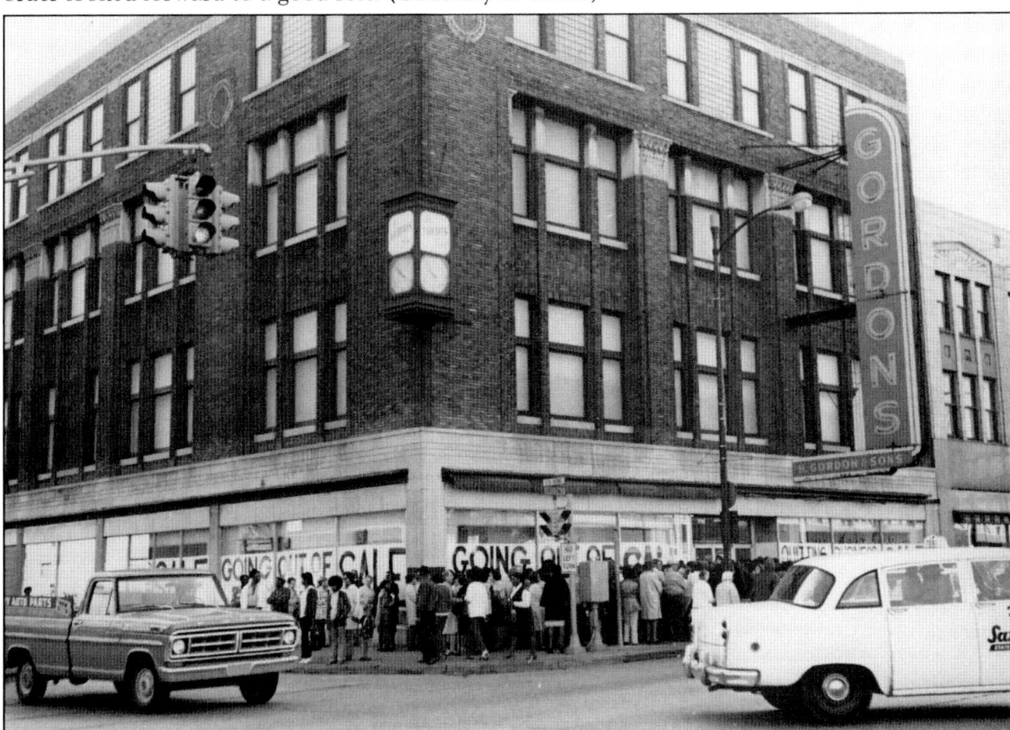

THE CLOSING OF H. GORDON & SONS, 1972. In the 1960s, many downtown Gary businesses were slowly leaving the city, some to suburban malls and others closing forever. Gordon's went out of business in 1972, but before that, it had a closing sale. People from throughout the area came back to the old store to make one last purchase, with a few there just to get an H. Gordon & Son store box. (Courtesy of CRA.)

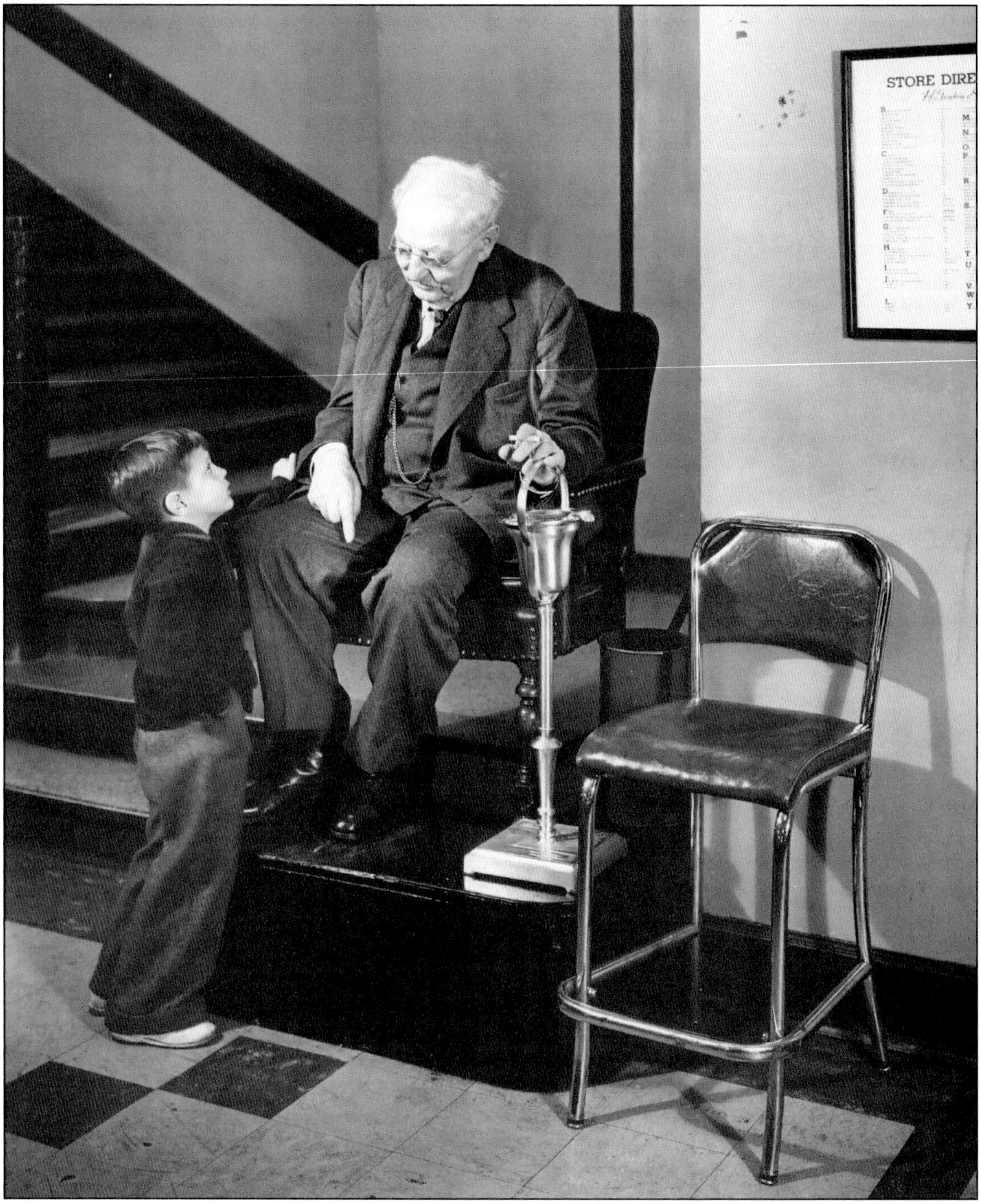

MR. H. GORDON. For years, this photograph of H. Gordon with a youngster was displayed near the elevators and main stairway. Many people still remember this portrait and have fond memories paired with it. (Courtesy of CRA.)

Five

HOUSES OF WORSHIP

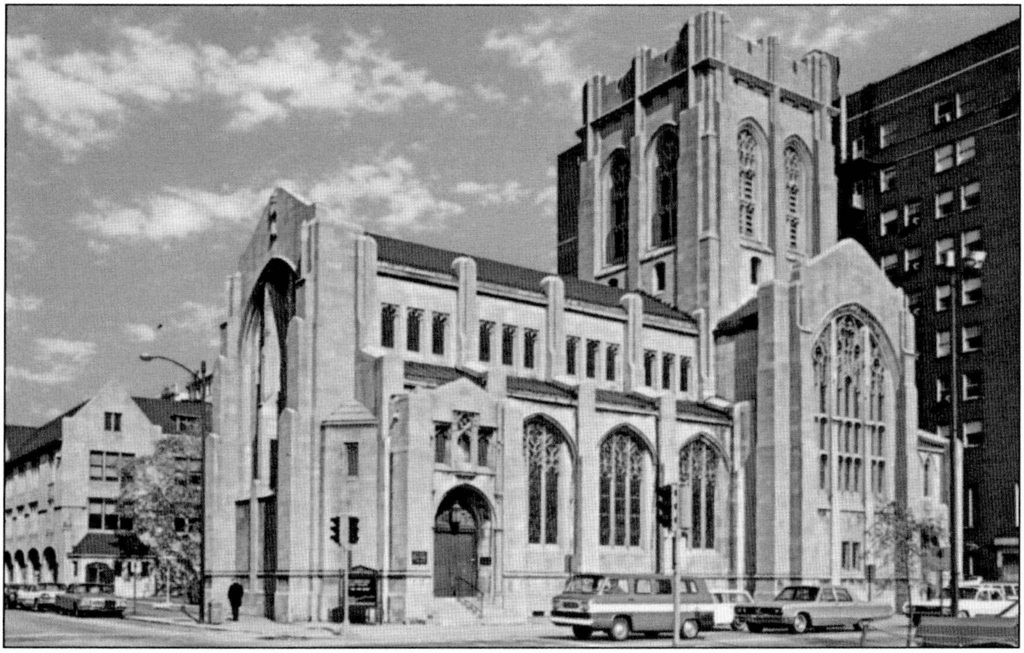

CITY METHODIST CHURCH. In 1926, the City Methodist Church was dedicated after nearly two years of work. The church, which stood at the northwest corner of Sixth Avenue and Washington Street, was the dream of Dr. William Grant Seaman. The Gothic-style structure contained the nave, educational classrooms, a hall, and a small commercial section. (Courtesy of CRA.)

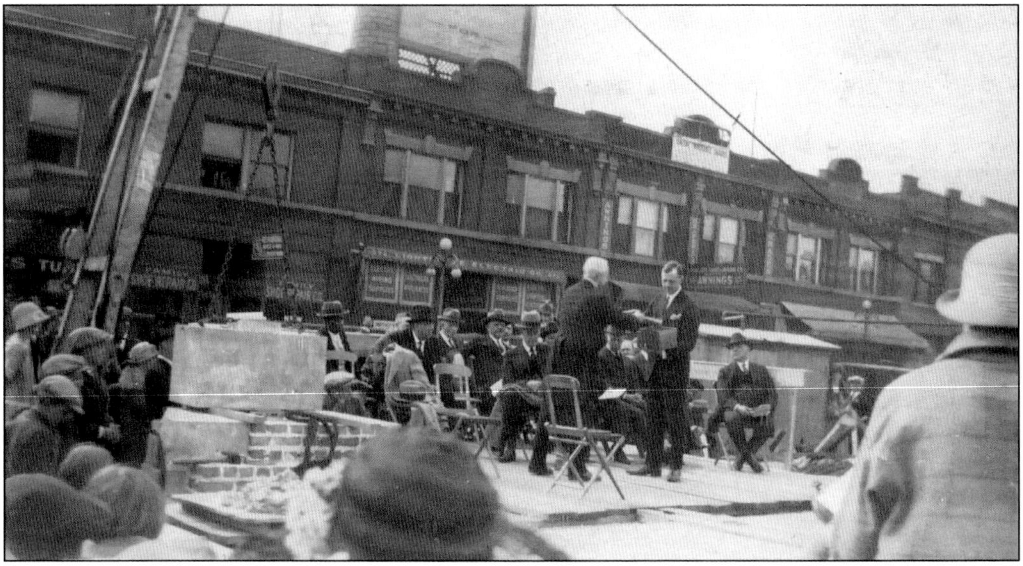

CITY METHODIST, CORNERSTONE. The City Methodist Church held a cornerstone-laying ceremony two years before the church was dedicated. In this photograph, Dr. William G. Seaman takes part in the event before a crowd of church members, dignitaries, and guests at Sixth Avenue and Washington Street. (Courtesy of CRA.)

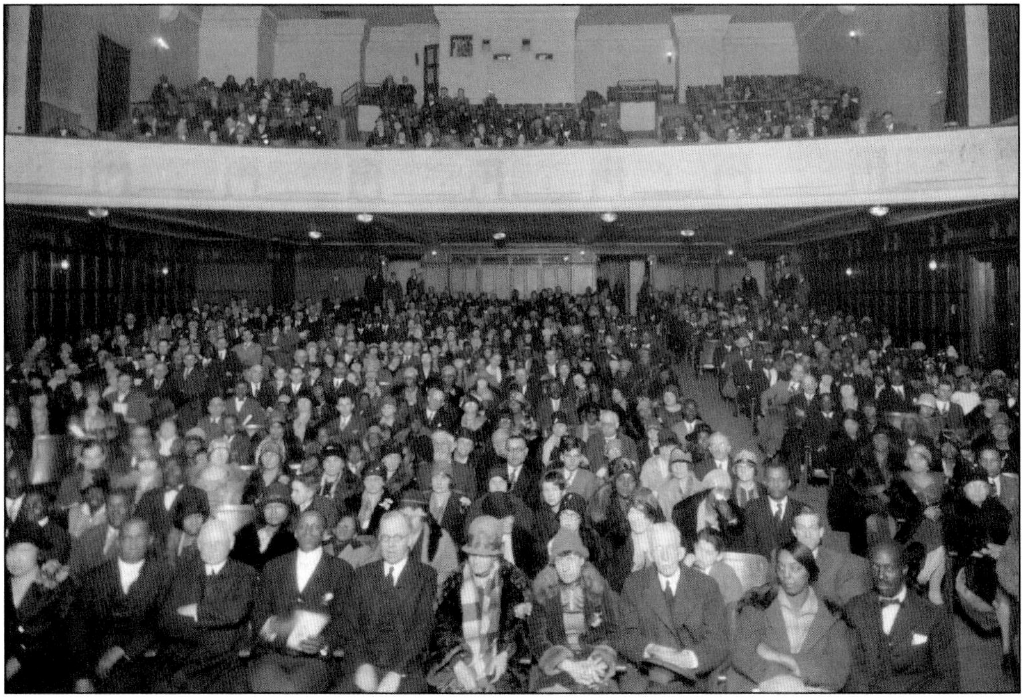

SEAMAN HALL. Seaman Hall stood just north of the main church. Named after the pastor, Dr. William Grant Seaman, the hall and theater area hosted church conferences, plays, and musical concerts. After World War II, Indiana University offered classes in the complex's classrooms. The classes were popular with many returning veterans. (Courtesy of CRA.)

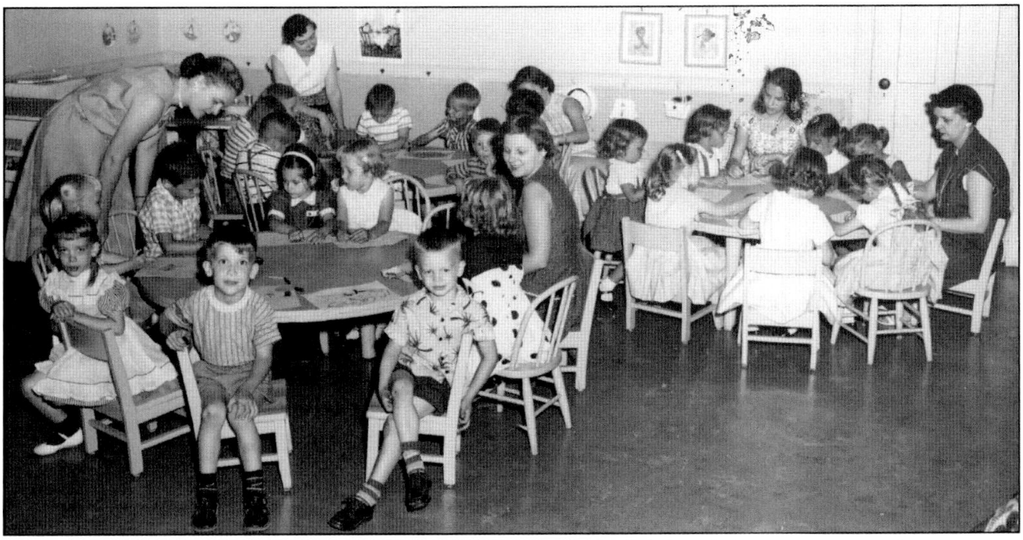

Sunday School, 1950s. The Methodist church included offices and classroom space for meetings and Sunday school. In this photograph taken in the early 1950s, children are shown working in groups on assigned projects. (Courtesy of CRA.)

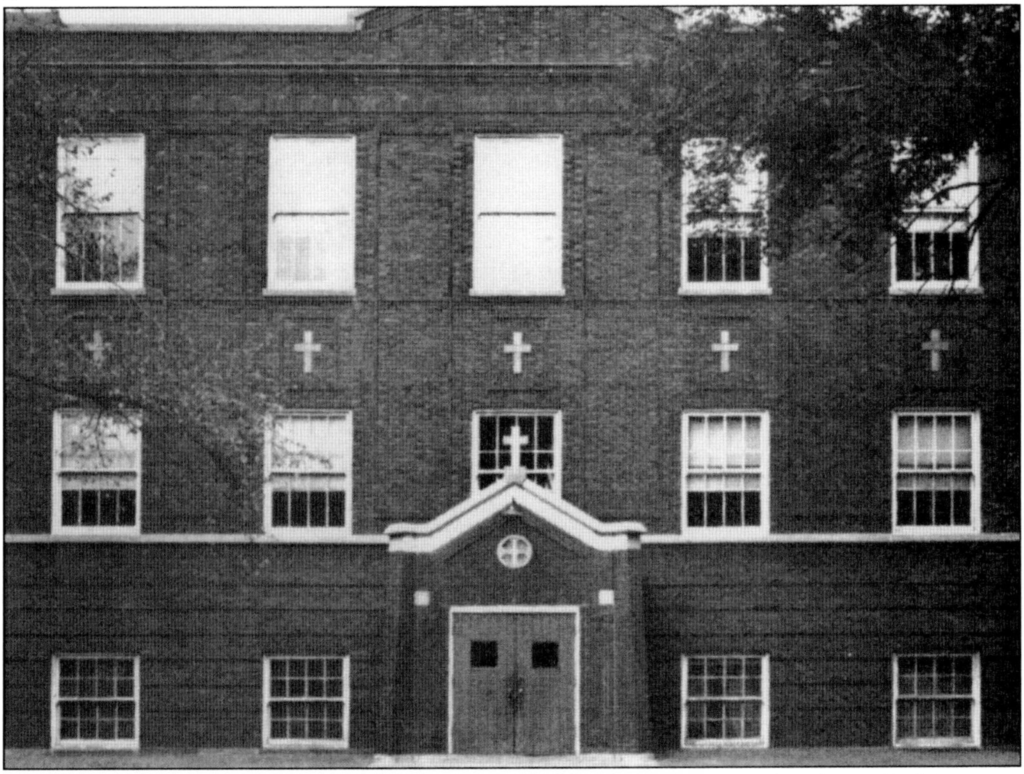

St. Luke Parish. In 1917, Holy Angel Parish was divided as growing membership called for a new church east of Broadway. Land was purchased along Seventh Avenue and Rhode Island Street, just east of Emerson High School. The bishop of the Diocese of Fort Wayne, Herman J. Alerding, assigned Rev. Frank J. Gnibba as its first pastor. That year, a church and school building were built. A rectory was also constructed in the 600 block of Vermont Street. (Courtesy of CRA.)

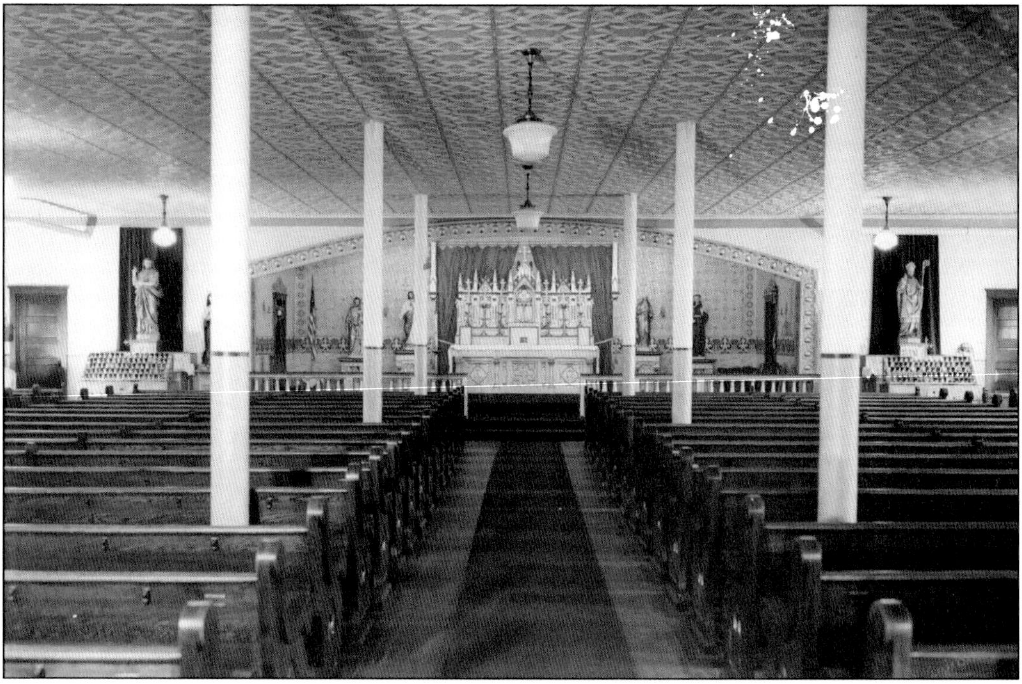

Original Church, 1930. St. Luke's original church was located on the lower level of the 1917 structure. When the present church was built, the lower level became a meeting hall for parish events. It is pictured here in the 1930s. (Courtesy of CRA.)

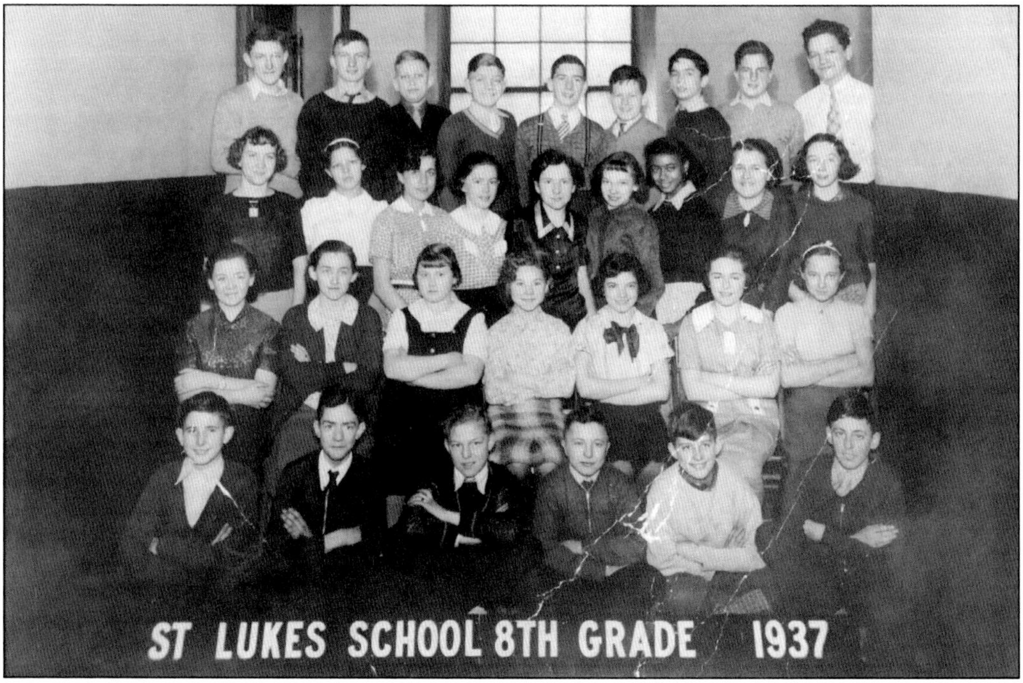

St. Luke's Eighth Grade, 1937. St. Luke School offered classes for students from kindergarten through the eighth grade. In the late 1930s, 500 families were members of the parish. Pictured is the eighth-grade class in 1937. (Courtesy of CRA.)

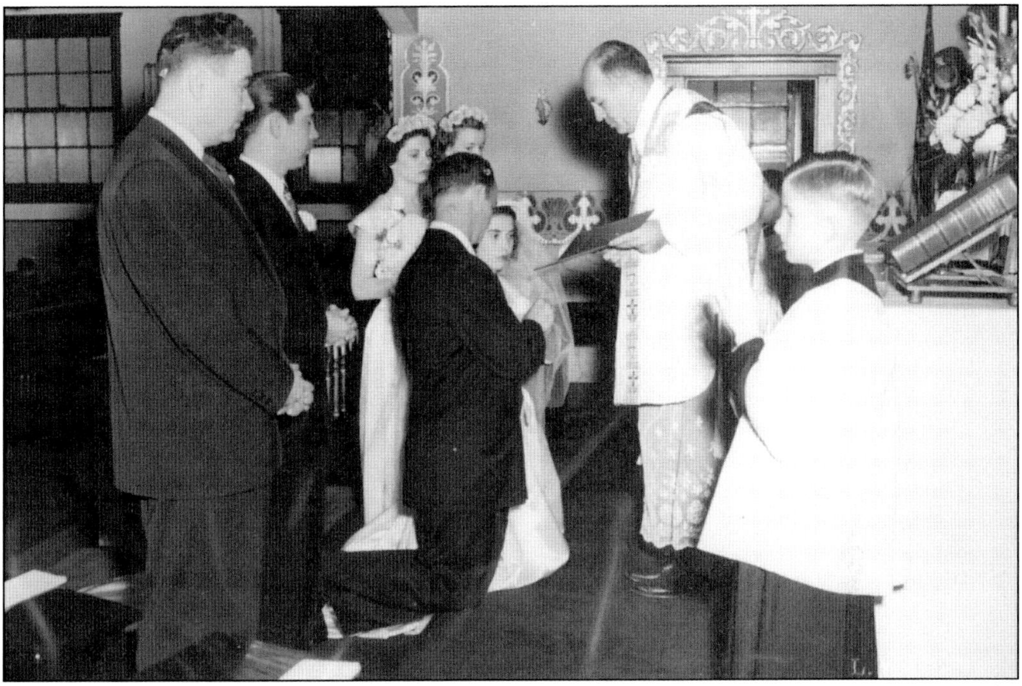

WEDDING MASS, 1945. When World War II came to an end, returning veterans got back to the lives they led before the conflict. In October 1945, decorated veteran Tom McMahon asked Mitzi Hunker to become his wife. The two exchanged vows at St. Luke Parish. (Courtesy of Colette Oram.)

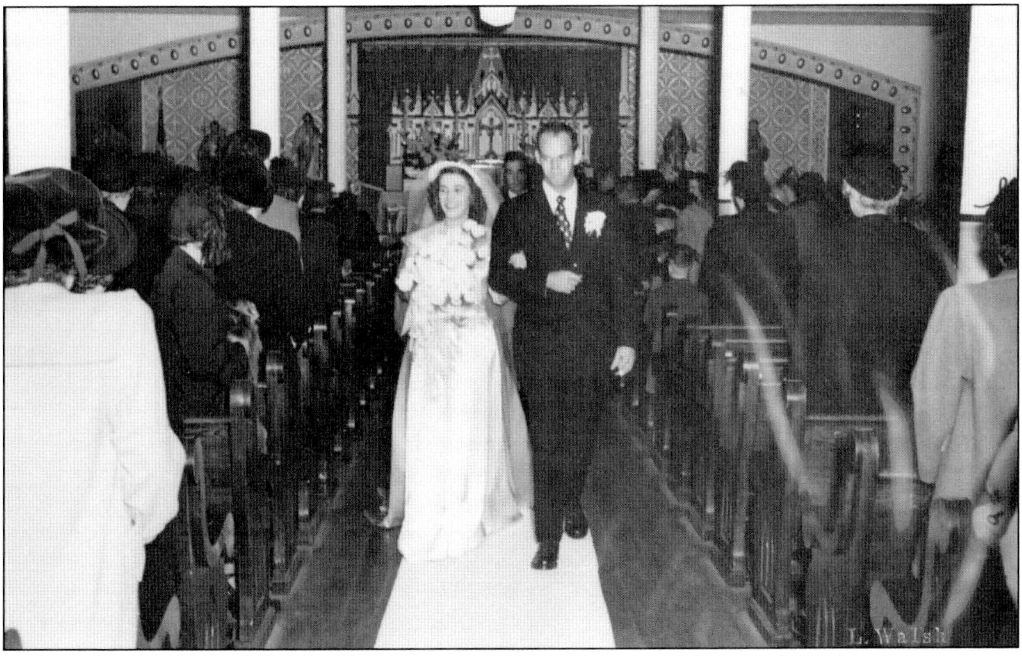

NEWLYWEDS TOM AND MITZI MCMAHON. Mr. and Mrs. McMahon depart St. Luke Church following their wedding Mass on October 14, 1945. The two were graduates of Emerson High School. Tom McMahon was awarded the Silver Star for gallantry while with the 1st Army in Europe. (Courtesy of Colette Oram.)

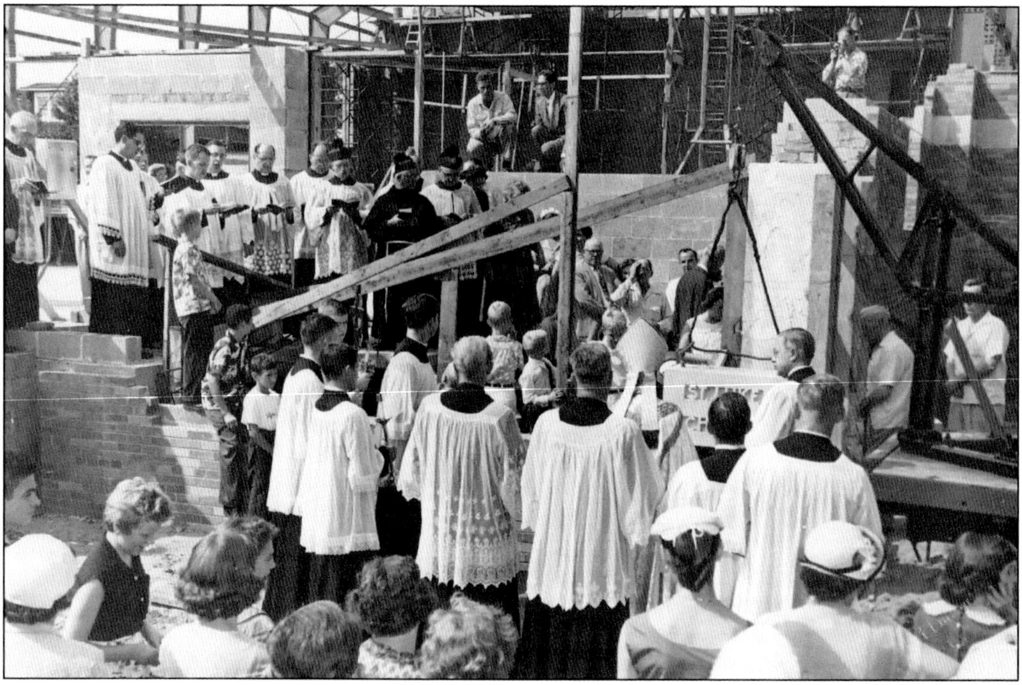

CORNERSTONE FOR NEW CHURCH, 1954. In the 1950s, plans were put together for the construction of a new church just east of the original building. Ground was broken in February 1954, and the cornerstone was laid that August. Pictured is the crowd that came for the blessing. (Courtesy of CRA.)

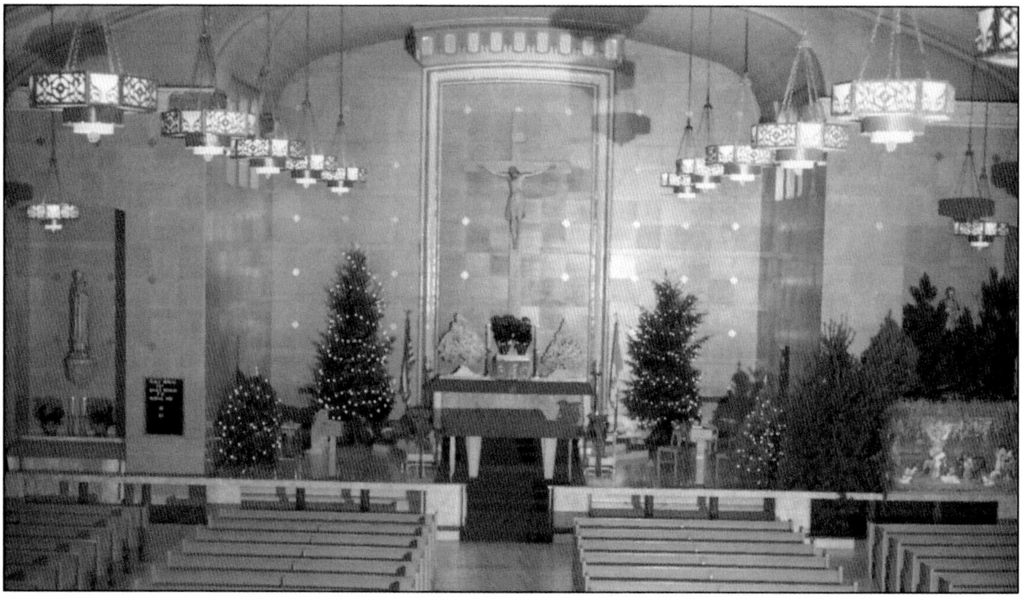

ST. LUKE CHURCH INTERIOR, 1960S. The new St. Luke Church was finally dedicated 16 months after the cornerstone was laid. On November 6, 1955, a Mass of dedication and thanksgiving was finally celebrated to a packed church. This mid-1960s photograph, taken from the choir loft, shows the church's interior during the Christmas season. After 63 years, the parish, now Ss. Monica and Luke, continues to serve the Catholic community of Gary's east side. (Courtesy of CRA.)

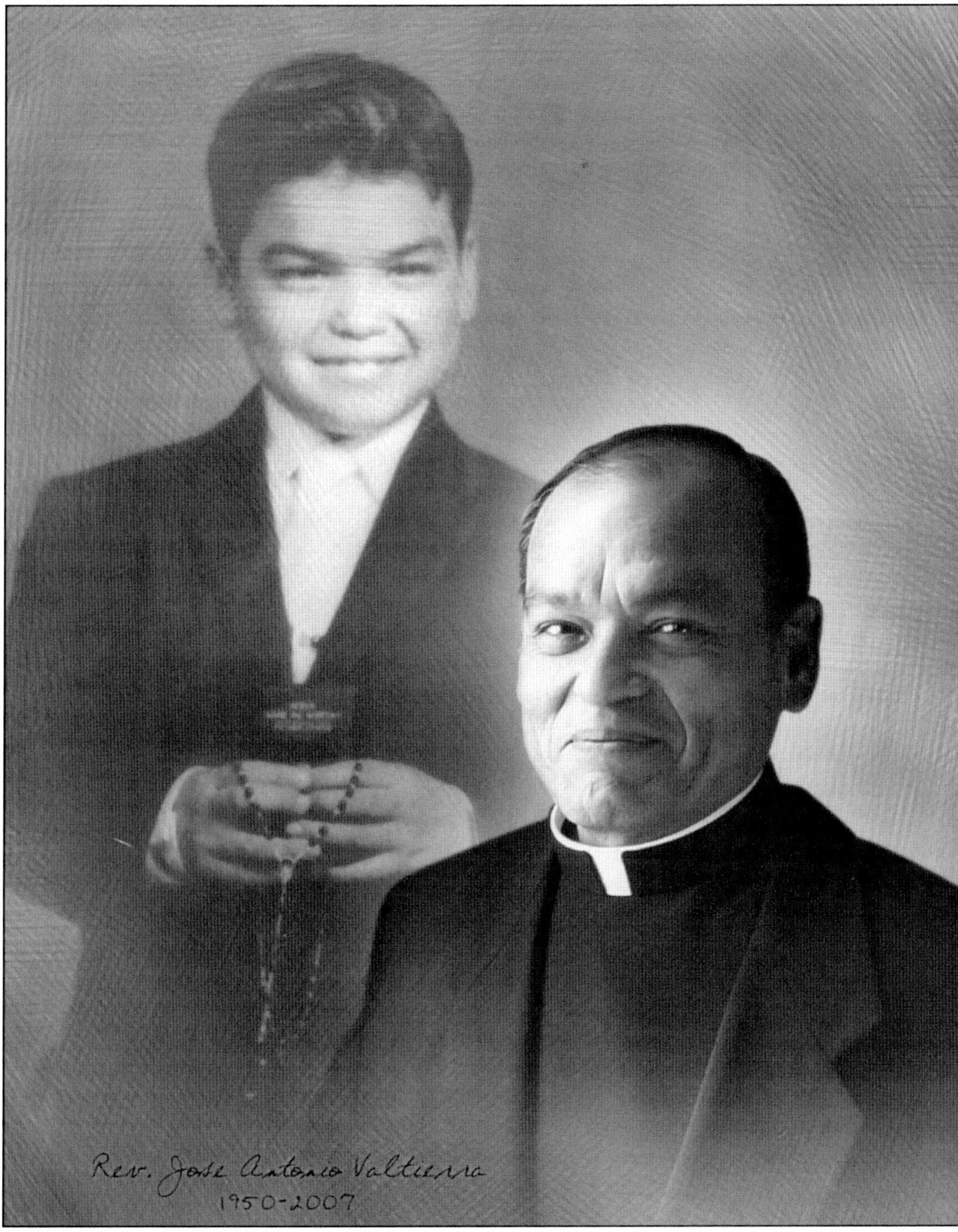

REV. TONY VALTIERRA. In 1979, Anthony Valtierra was ordained a priest and served the Diocese of Gary. He was a member of St. Luke Parish and a 1968 graduate of Emerson High School. Father Valtierra studied at St. Meinrad Seminary in Indiana and was at St. Francis Parish before his passing in 2005. (Courtesy of Michael Valtierra and family.)

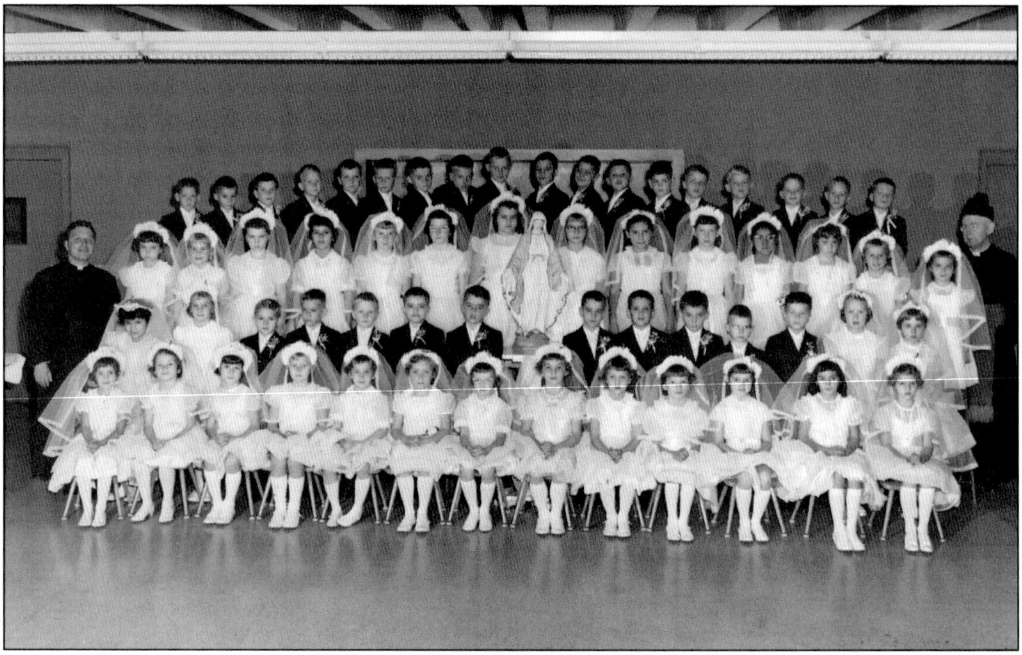

First Communion Class. Pictured is the St. Luke First Communion class of 1960. Fr. William Martin is on the left, and the church's pastor, Fr. H. James Conway, is on the right. (Courtesy of Colette Oram.)

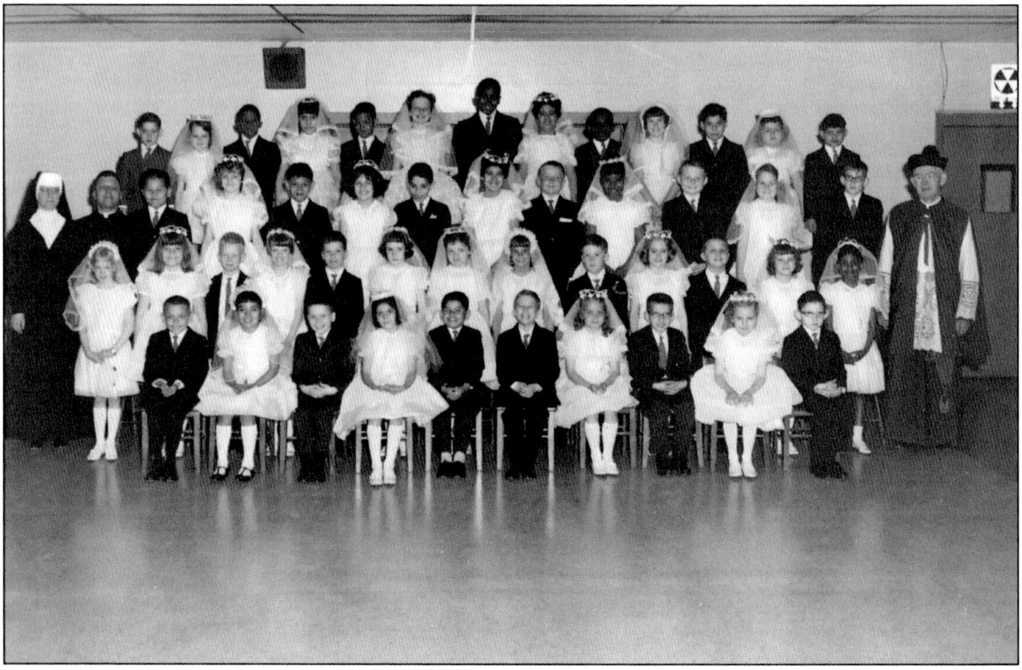

First Communion Class, 1966. The St. Luke First Communion class of 1966 is shown here. Sister Katherine and Father Martin are on the left, and Father Conway is on the right. (Courtesy of Diane Trafny.)

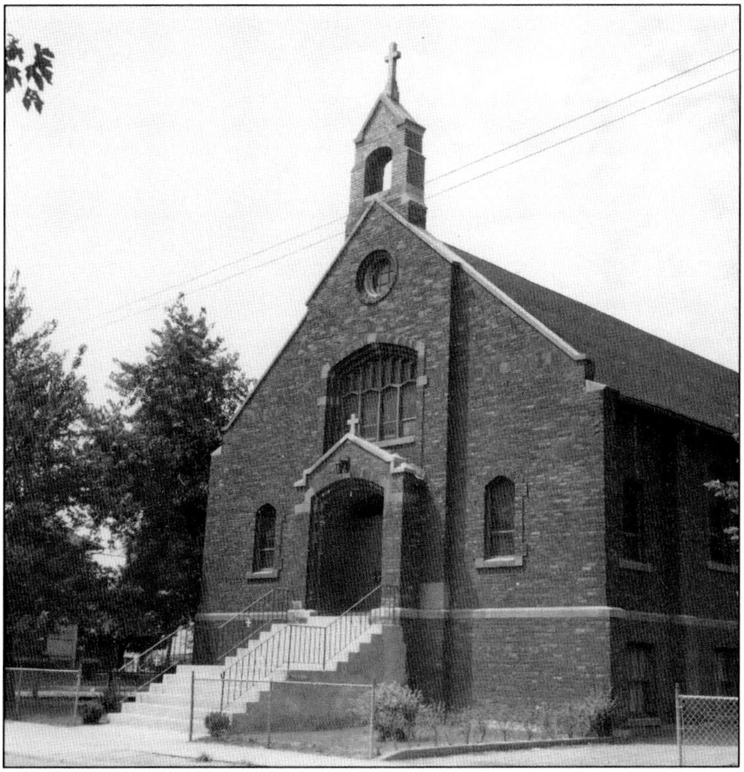

St. Monica Church, 1940s. The original parish of St. Monica was located at Twenty-Second Avenue and Adams Street. It saw to the spiritual needs of the African American Catholics in Gary's Central District. The church is pictured here in the late 1940s. (Courtesy of Ss. Monica and Luke Parish.)

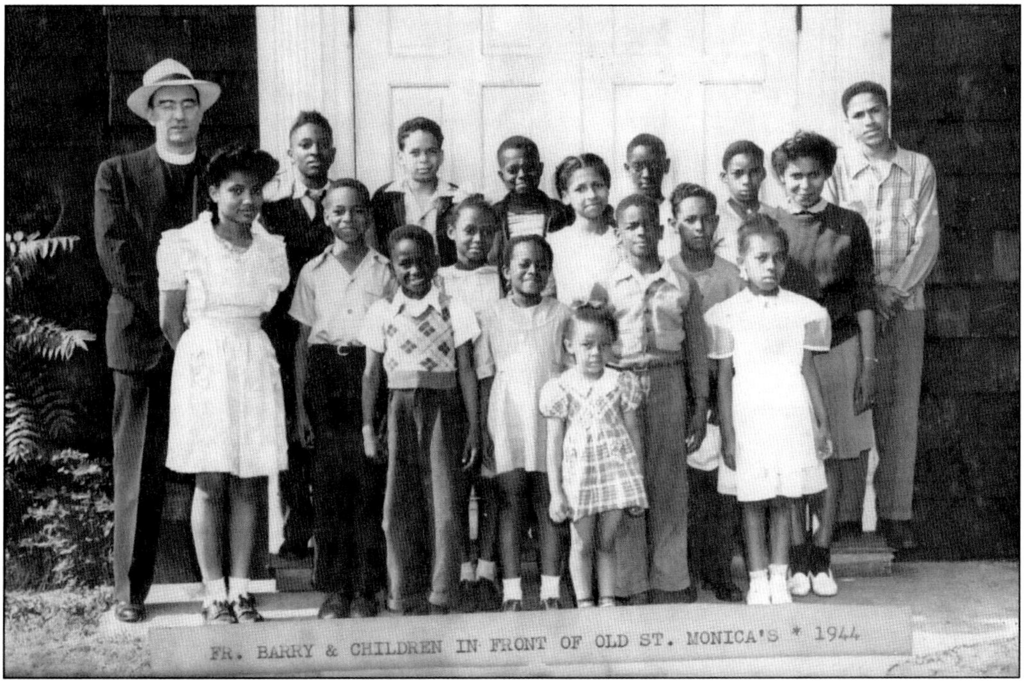

St. Monica Students, 1944. St. Monica students are pictured with their pastor, Father Barry, in front of the original church building in 1944. A brick church was constructed at Twenty-Second Avenue that replaced the original building. (Courtesy of Ss. Monica and Luke Parish.)

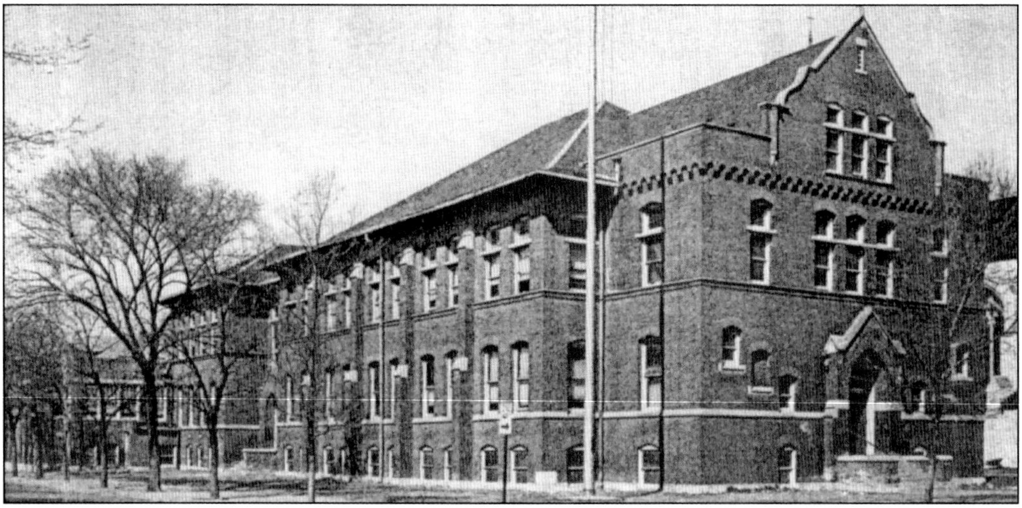

Holy Angels Parish, 1950. Holy Angels Parish was established in 1909 to serve the growing Catholic community of Gary's north side and downtown area. A church and school building were used until 1950, when a new church was built. Seven years later, it became a cathedral when the Diocese of Gary was established. The old building is shown here around 1950. (Courtesy of CRA.)

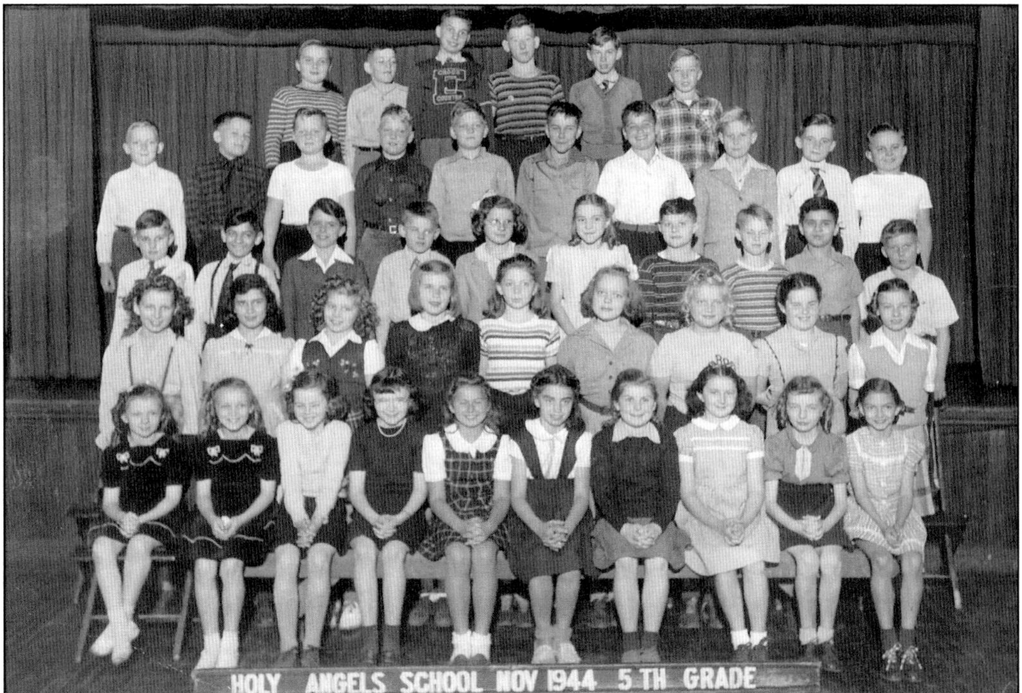

Holy Angels Fifth Grade, 1944. During World War II, American life went on at the home front but with restrictions. Rationing, bond sales, censored mail, and blackouts along the coasts were common. Sacrifice and prayers were offered for those overseas. Pictured is the fifth-grade class of Holy Angels in November 1944. It is quite possible that a loved one of this group of students would be involved in the coming Battle of the Bulge that winter, the biggest battle of the war in Western Europe. (Courtesy of Tom Campbell.)

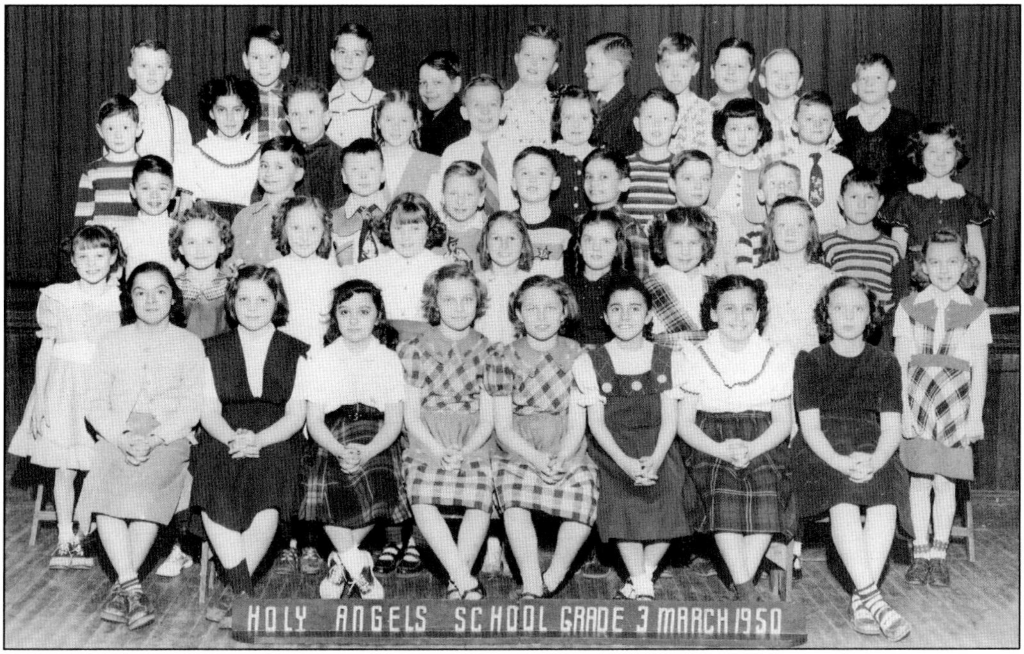

HOLY ANGELS THIRD GRADE, 1950. Pictured is the third grade of Holy Angels School in 1950. Though considered part of the World War II generation, they are not considered part of the huge baby boom group as they were born before 1946. (Courtesy of Don Mueller.)

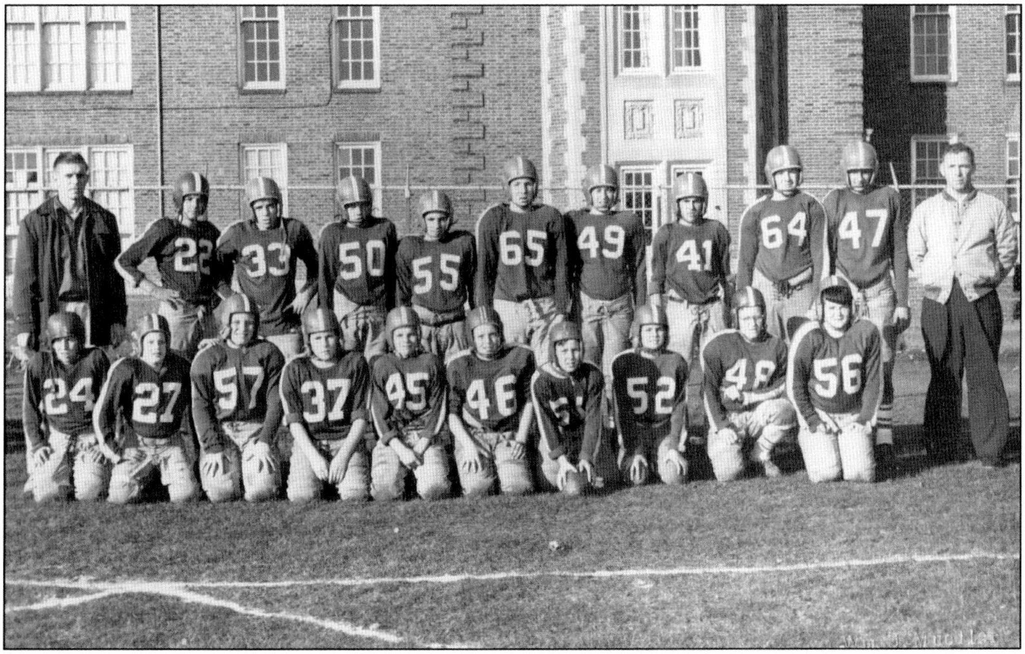

CATHOLIC YOUTH ORGANIZATION FOOTBALL, 1950S. The Catholic Youth Organization helped sponsor a number of sports teams for boys and girls in the Diocese of Gary. Football for the boys was popular with many Catholic grade schools in Gary, and both St. Luke and Holy Angels fielded teams for many years. Pictured are the Holy Angels Bears in 1955 at Tolleston Field. (Courtesy of Don Mueller.)

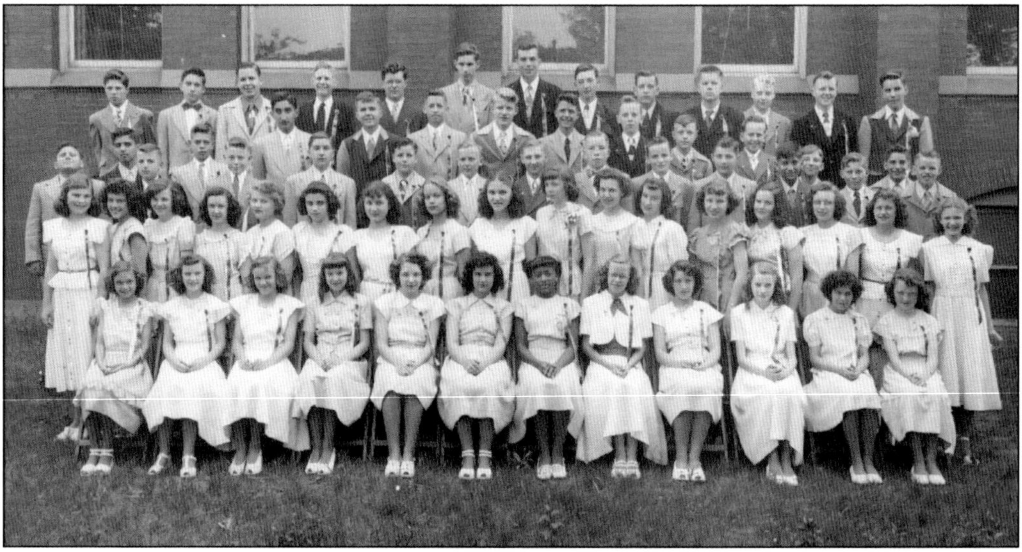

HOLY ANGELS GRADUATION, JUNE 1948. Pictured near the parish school is the 1948 graduating class of Holy Angels. The young graduates were not only thankful for their academic success but also for the safe return of loved ones from combat in Europe or the Pacific. (Courtesy of CRA.)

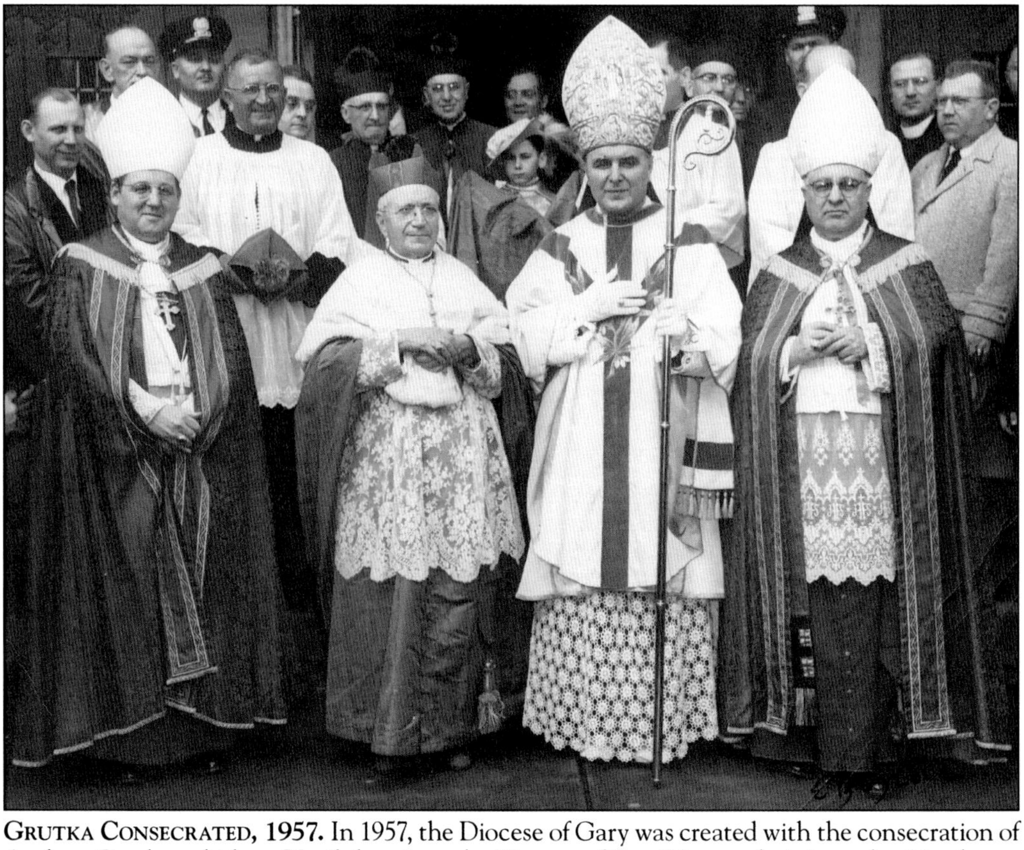

GRUTKA CONSECRATED, 1957. In 1957, the Diocese of Gary was created with the consecration of Andrew Grutka as bishop. Until that time, the Diocese of Fort Wayne administered to Northwest Indiana. Bishop Grutka is shown here before Holy Angels Cathedral. (Courtesy of CRA.)

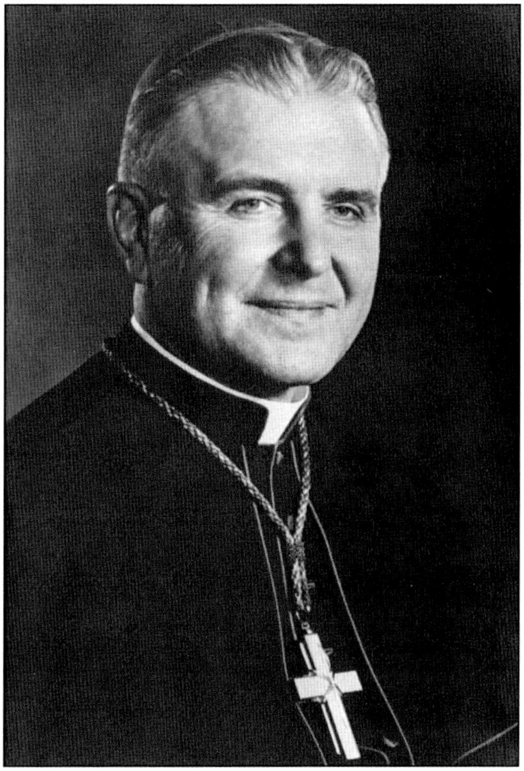

Bishop Andrew Grutka. Bishop Andrew Grutka was born in Joliet, Illinois, and after his ordination, his parish assignment sent him to Holy Trinity in Gary. In 1957, he became bishop of the new Diocese of Gary where he served for 28 years. He retired in 1984. (Courtesy of CRA.)

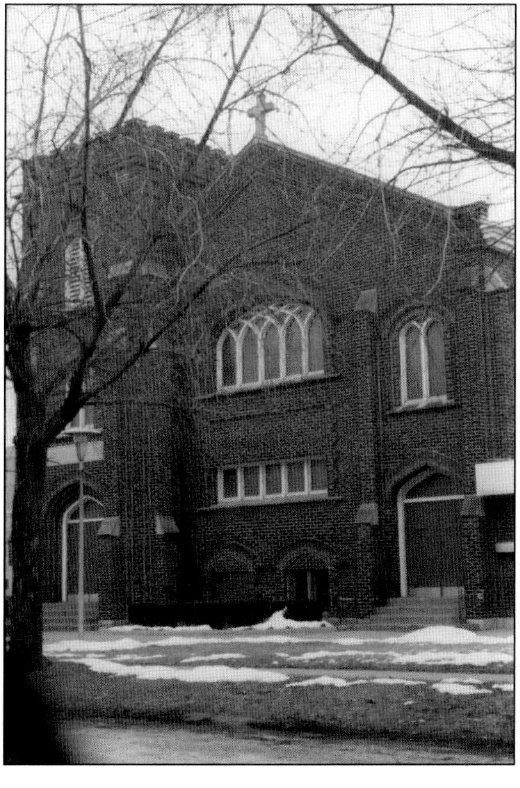

Bethlehem Lutheran. Bethlehem Lutheran Church was built on the southeast corner of Sixth Avenue and Fillmore Street and still holds services for its west side congregation. It is pictured here around 1980. (Courtesy of CRA.)

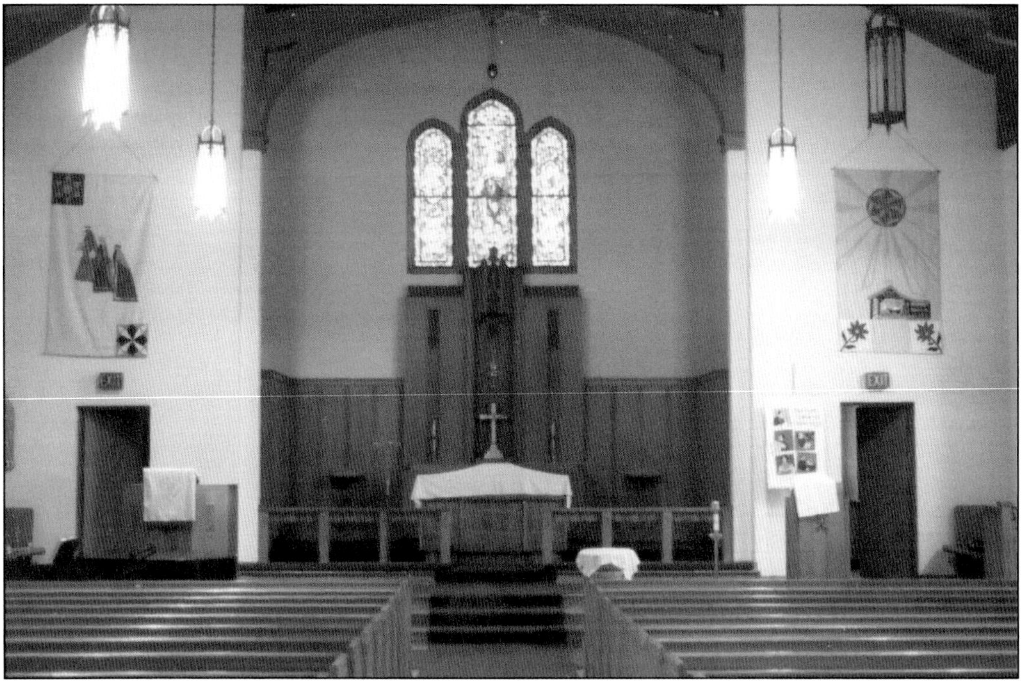

Bethlehem Lutheran Interior. The interior of Bethlehem is shown here around 1990. (Courtesy of CRA.)

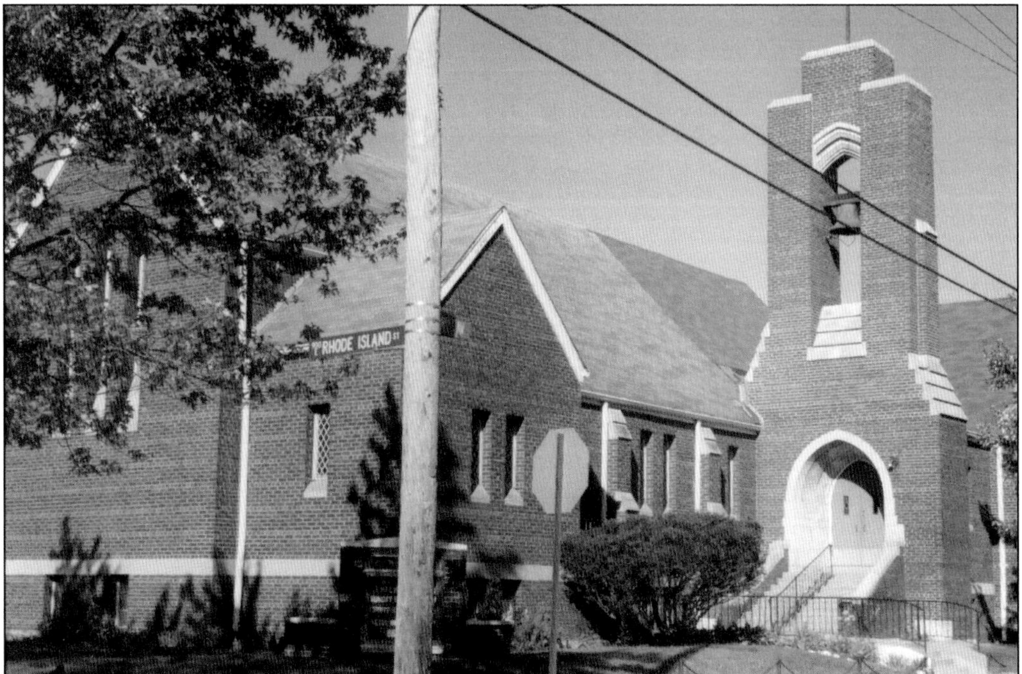

Centennial United Methodist Church. Pictured is the Centennial United Methodist Church at Eighth Avenue and Rhode Island Street in the 1980s. Earlier, it was the United Brethren Church. (Courtesy of CRA.)

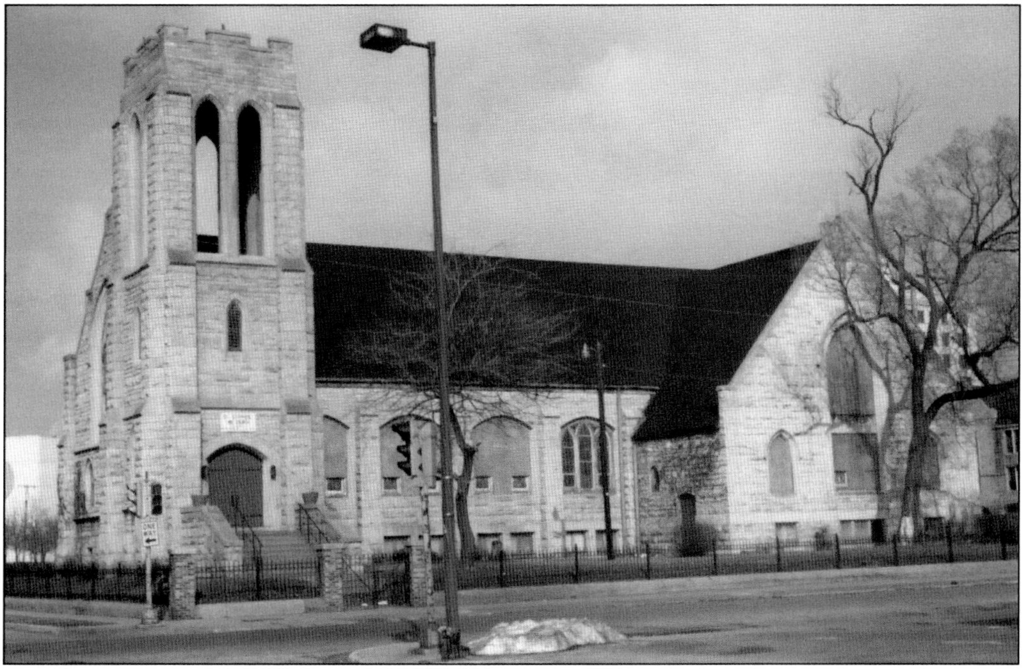

Christ Church. Along west Sixth Avenue is Christ Episcopal Church just a block away from the 21st Century Charter School. In this photograph, the sign above the main entrance reads St. Stephen Missionary Baptist. (Courtesy of CRA.)

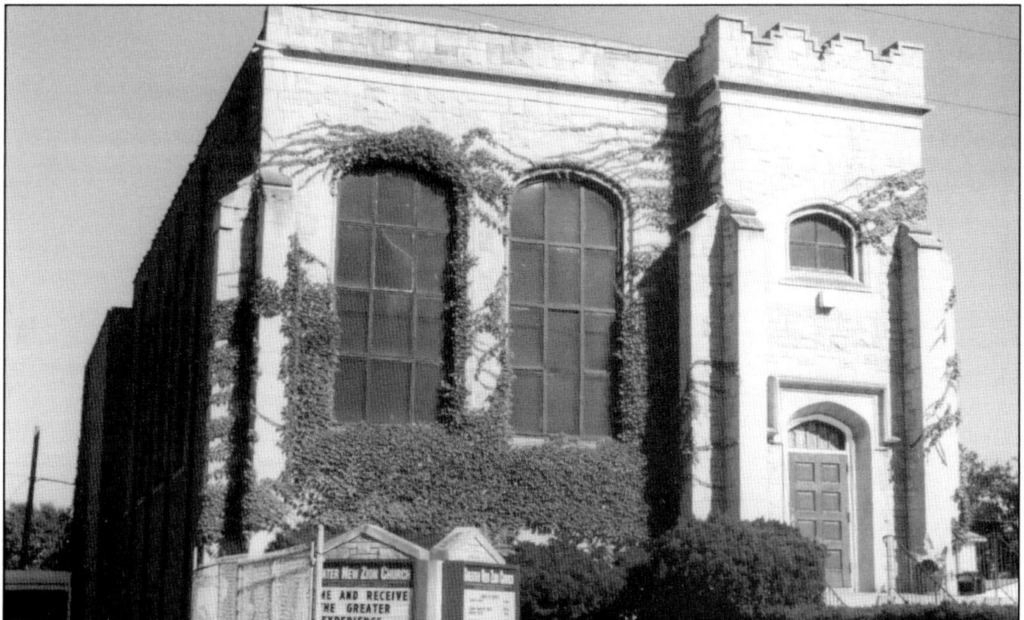

First Congregational Church. Gary's First Congregational Church, located in the 600 block of Madison Street, was constructed during the Steel City's early years. The building, which has continued a spiritual mission to the community for a century, is now the River of Life Church. (Courtesy of CRA.)

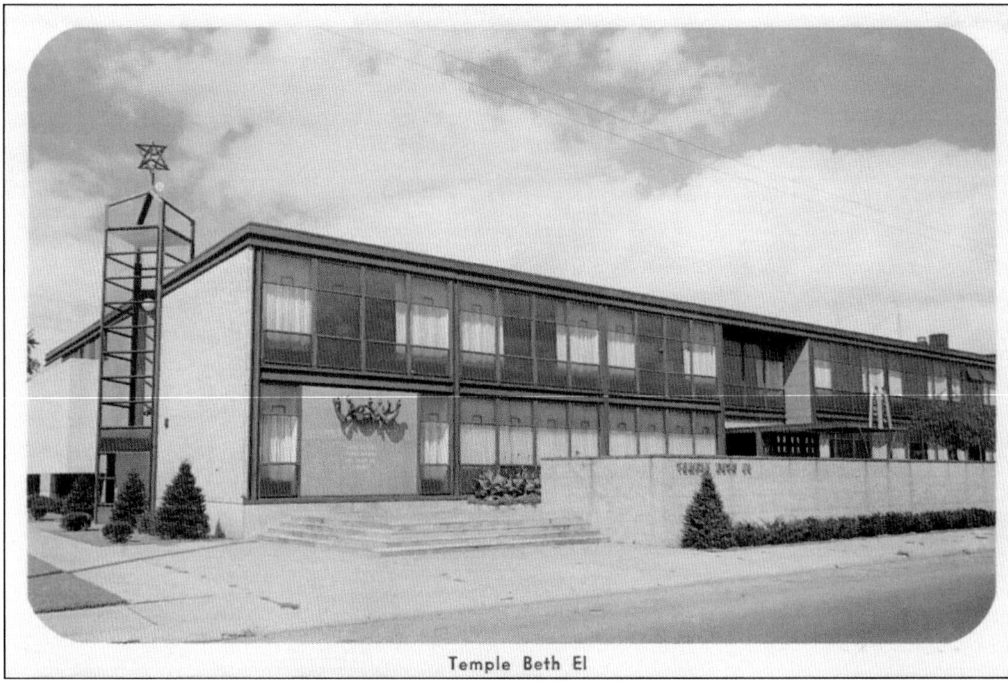

Temple Beth El. Jewish families from the Horace Mann community attended worship services at the Temple Beth El for many years. The synagogue was located near the 1800 block of West Fifth Avenue. The congregation has since moved to Munster. (Courtesy of CRA.)

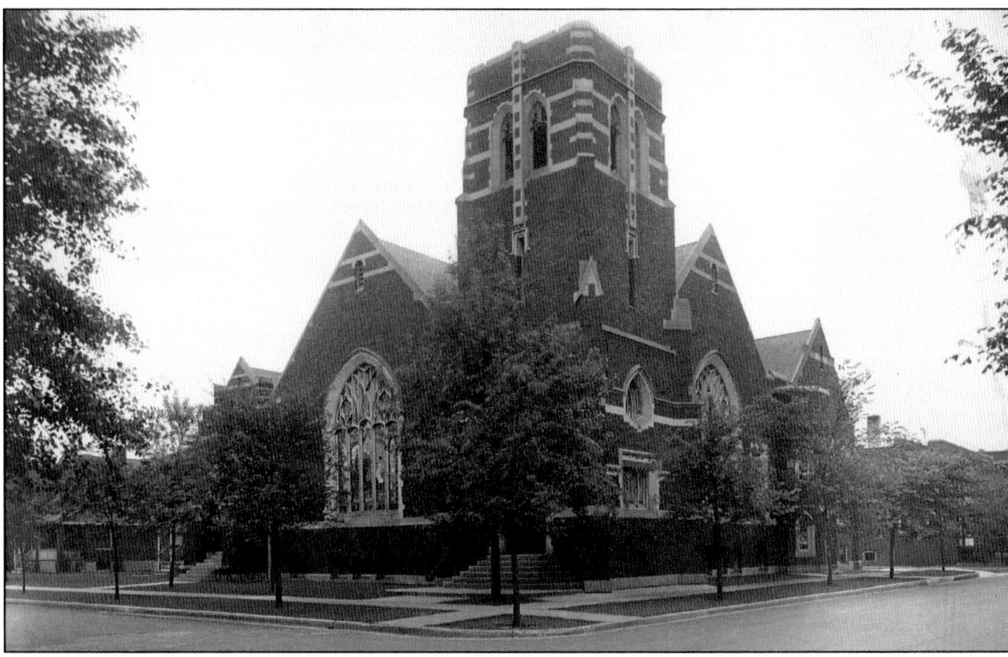

First Presbyterian, 1940. Gary's First Presbyterian Church, located at 574 Monroe Street, was dedicated in 1913. The building suffered fire damage in 1947 in the main altar area, but the damage was soon repaired. In 2005, the congregation voted to sell the building. It is pictured here in 1940. (Courtesy of CRA.)

First Presbyterian Interior, 1950. The interior of First Presbyterian is shown in this c. 1950 photograph. Parts of the main altar and worship area were damaged but quickly restored after the fire in 1947. (Courtesy of CRA.)

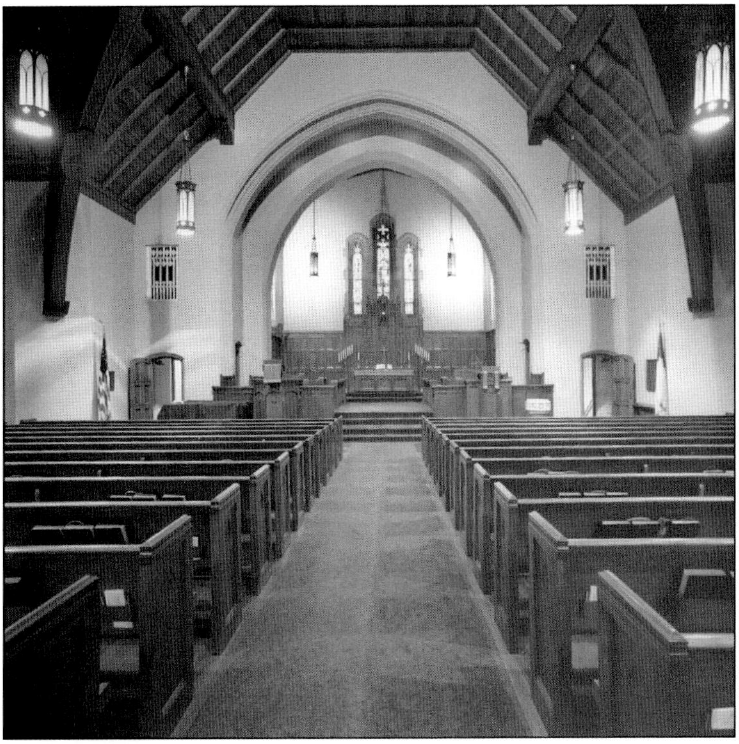

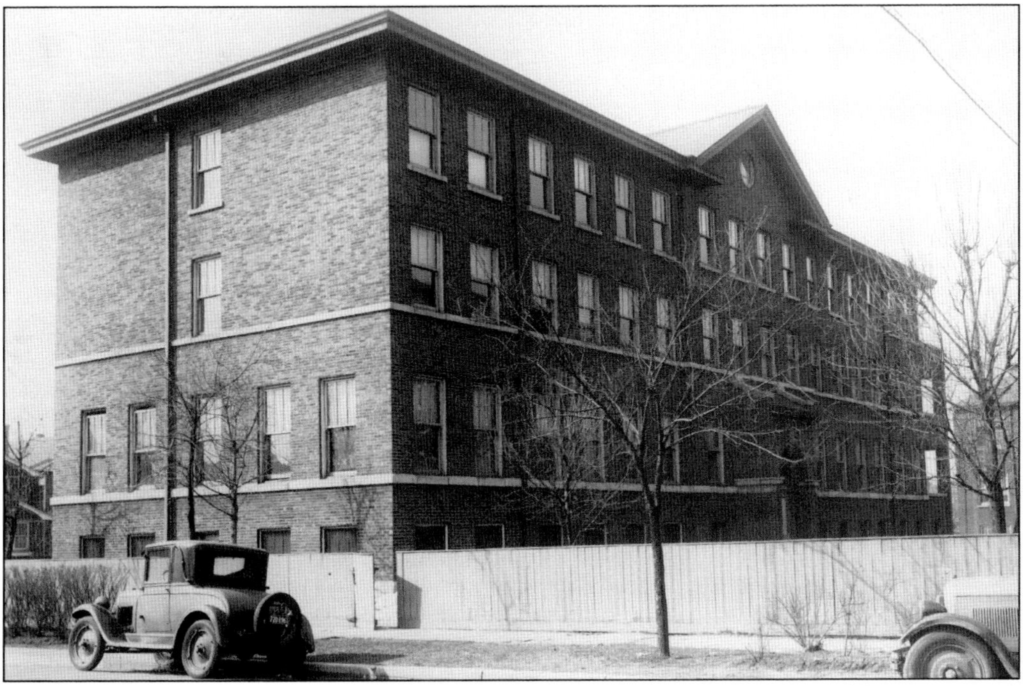

Nursing School and Home, 1922. Mercy Hospital moved from its original building at Seventh Avenue and Carolina Street in 1913. The new facility included the main hospital building, an operating space, and nursing school with dormitory. Pictured is the original school that was run by the sisters of the Poor Handmaids of Jesus Christ. (Courtesy of CRA.)

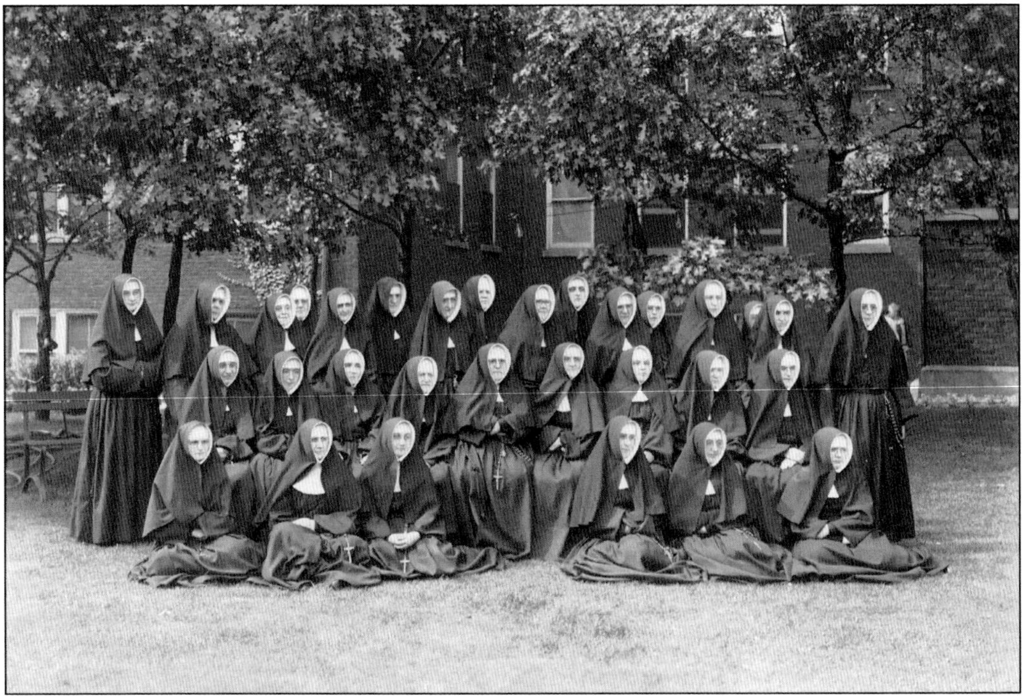

POOR HANDMAIDS OF JESUS CHRIST, 1923. In 1913, the Poor Handmaids of Jesus Christ came to Mercy Hospital as part of their religious mission to minister to the sick. The order, which began in Europe in 1851, later expanded to the United States and went on to work as nurses, hospital administrators, and nursing instructors. (Courtesy of CRA.)

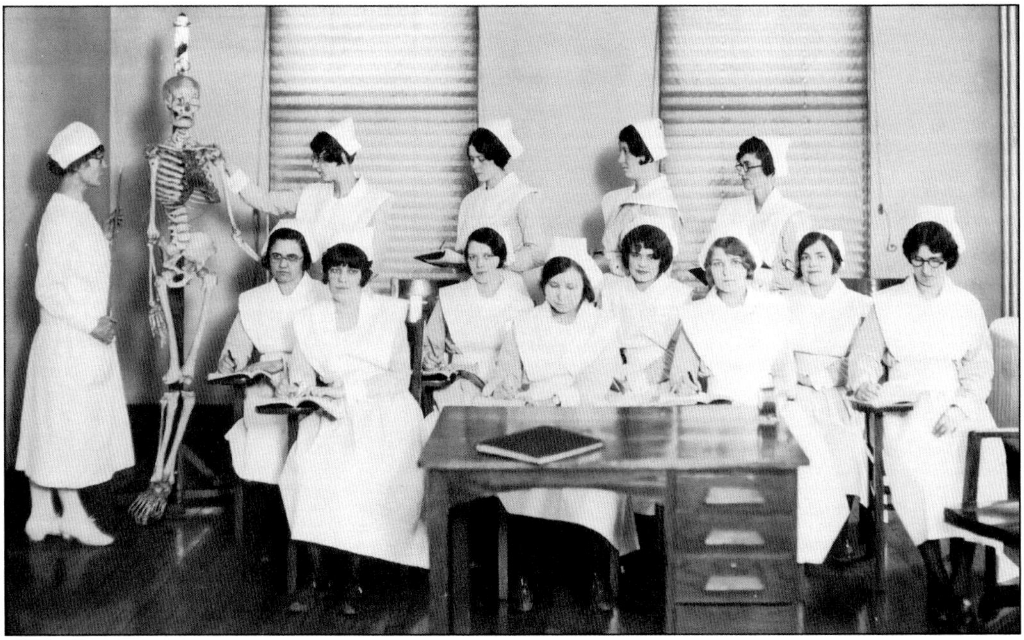

NURSING SCHOOL, 1929. Mercy Hospital of Gary operated a nursing school at its west side campus as far back as the early 1920s. Pictured is an anatomy class for young female students in 1922. The skeleton at left was not part of the faculty. (Courtesy of CRA.)

Six

Emerson and Horace Mann, Longtime Rivals

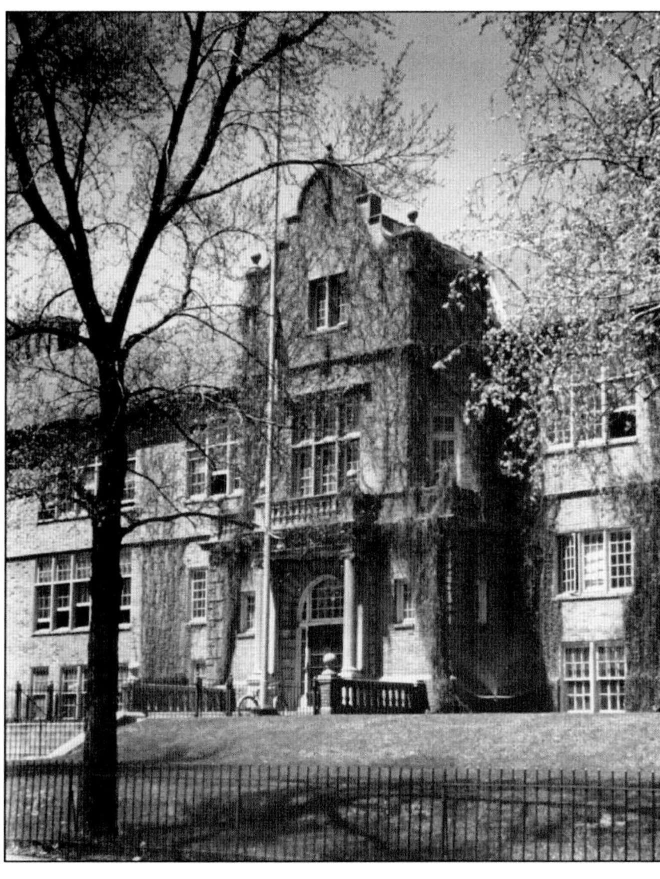

Emerson School, 1950. In 1907, construction began on Emerson School at the northeast corner of Seventh Avenue and Carolina Street. When Emerson opened, it became a modern school under the direction of Dr. William Wirt. The school had an eight-hour day, rotating classes, labs, gyms, an auditorium, shops for boys, band, and domestic science for girls. In addition, the school had its own paper, photography department and darkroom, small zoo, and gardens. The Wirt plan was based on work, study, and play. Work included shop or job training, study called for academics, and play involved sports, gym, and recess for elementary school. In the late 1920s, Emerson began a rivalry with Horace Mann that would last for over 40 years. (Courtesy of CRA.)

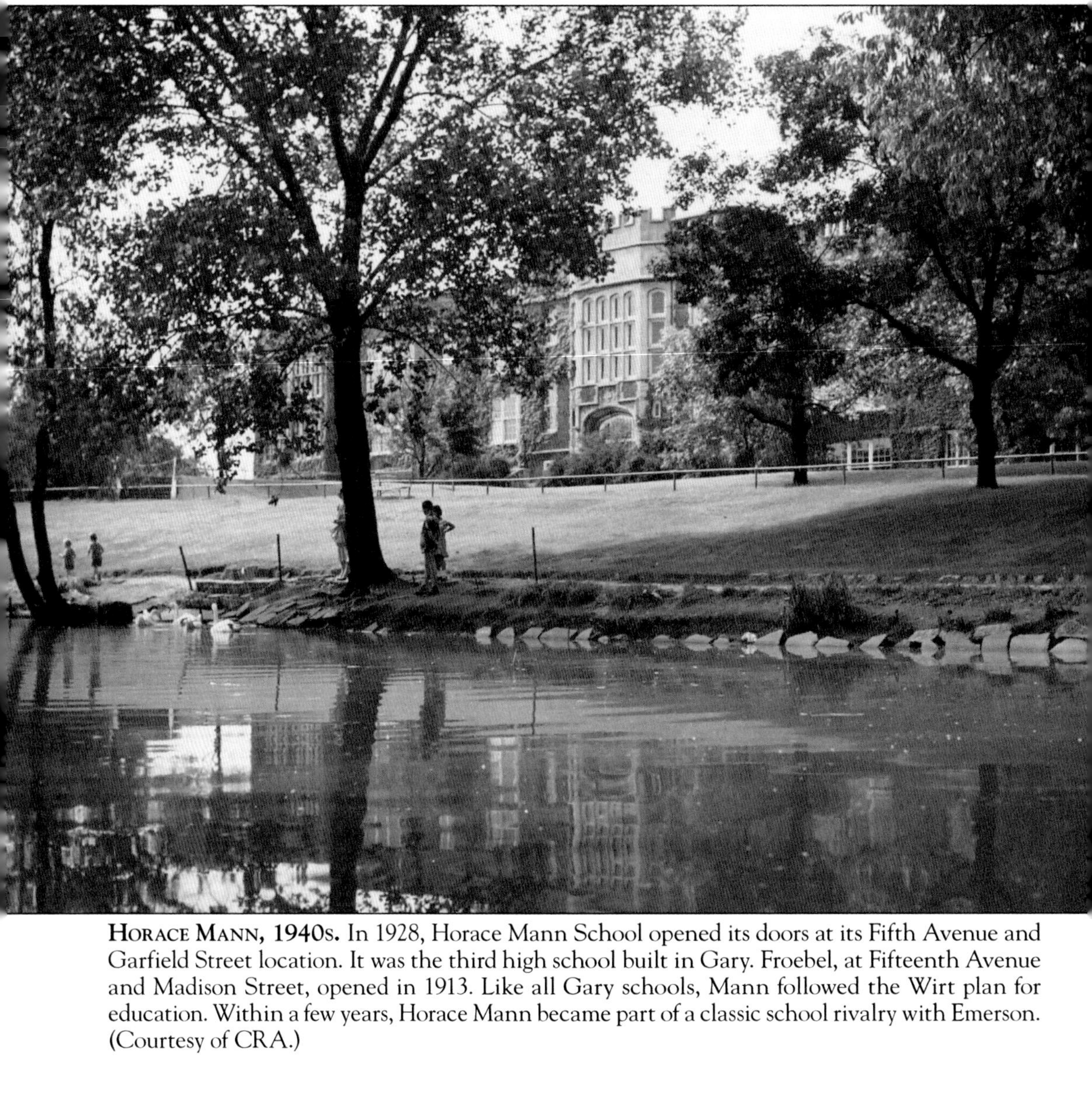

Horace Mann, 1940s. In 1928, Horace Mann School opened its doors at its Fifth Avenue and Garfield Street location. It was the third high school built in Gary. Froebel, at Fifteenth Avenue and Madison Street, opened in 1913. Like all Gary schools, Mann followed the Wirt plan for education. Within a few years, Horace Mann became part of a classic school rivalry with Emerson. (Courtesy of CRA.)

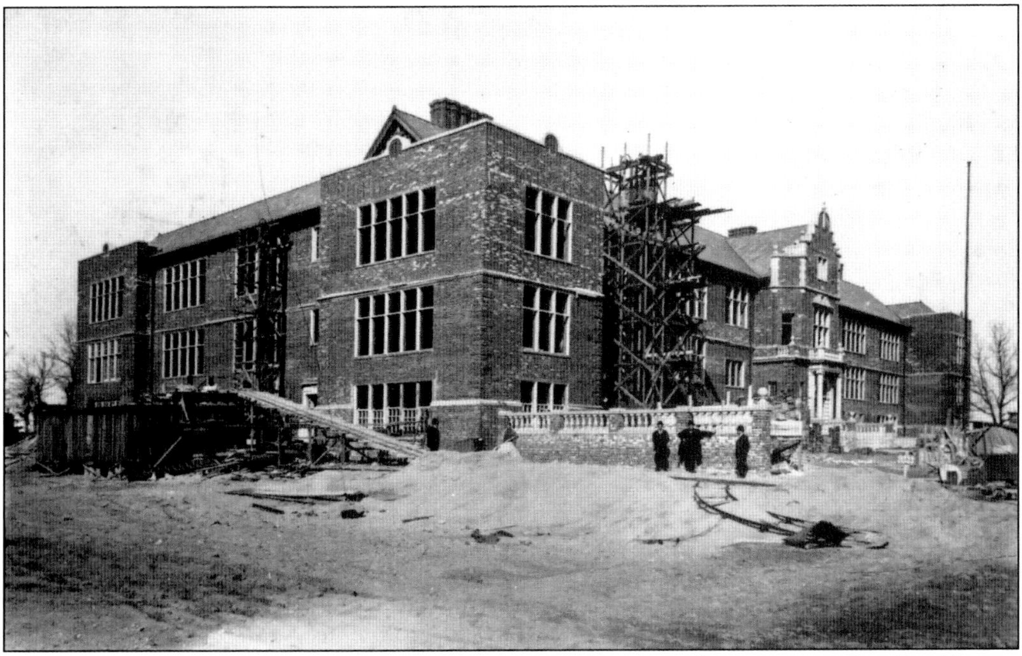

Emerson Construction, 1906. In 1906, construction began on the Emerson School building at Seventh Avenue. Considering it was done in the early 1900s, construction progressed at a rather brisk pace. In this photograph, the outside work is nearly complete. (Courtesy of CRA.)

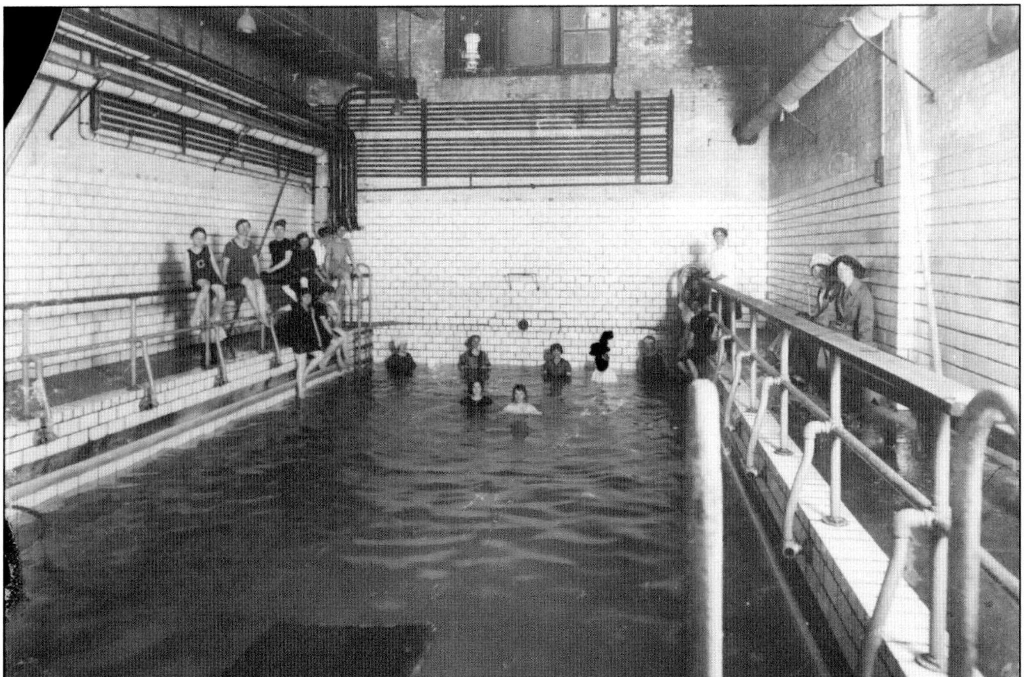

Emerson Pool, 1914. Emerson was one of the first high schools in the Midwest that had an indoor pool for student use in gym and competitions. Though quite small compared to the Olympic-size pools in modern high schools, it was a first for education. (Courtesy of CRA.)

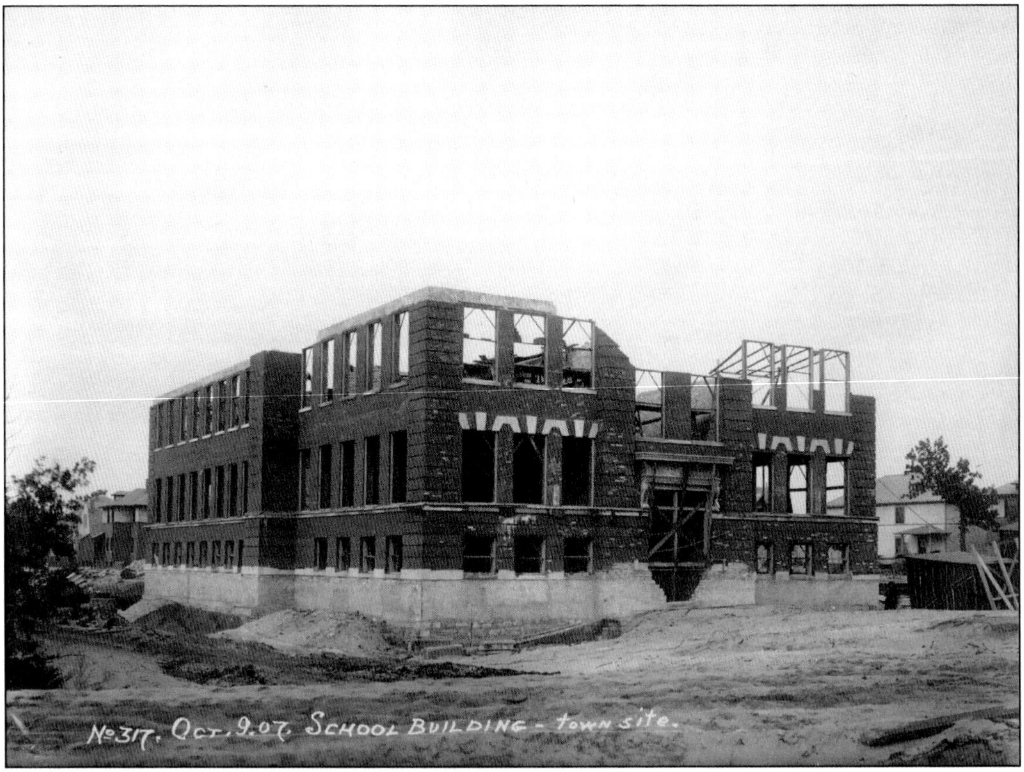

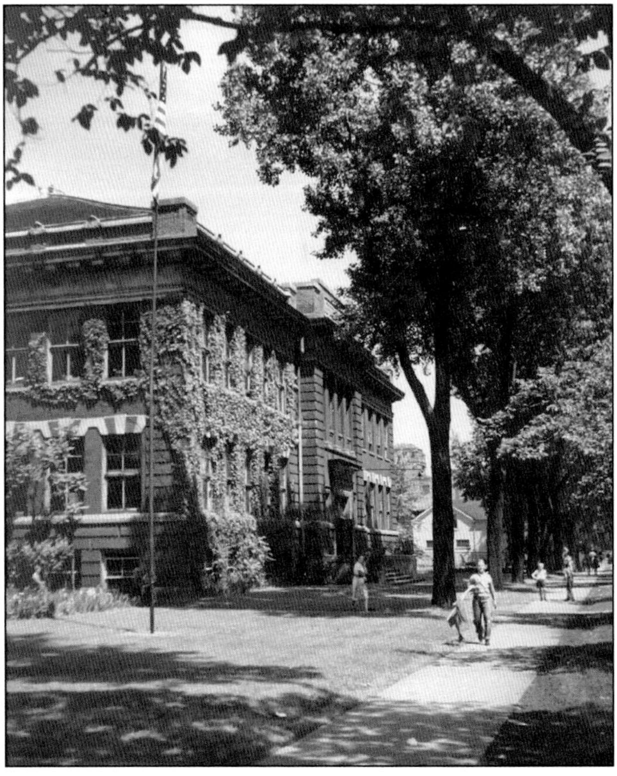

Jefferson School Construction, 1907. As Gary's school-age population continued to grow, the small school along Fourth Avenue downtown proved too small. In 1907, Jefferson School at West Seventh Avenue was constructed. The population had grown to nearly 500 children at the building's opening. In two years, Emerson School opened its doors. (Courtesy of CRA.)

Jefferson School, 1936. Pictured is the completed Jefferson School in the spring of 1936. The building served as an elementary school at that time. (Courtesy of CRA.)

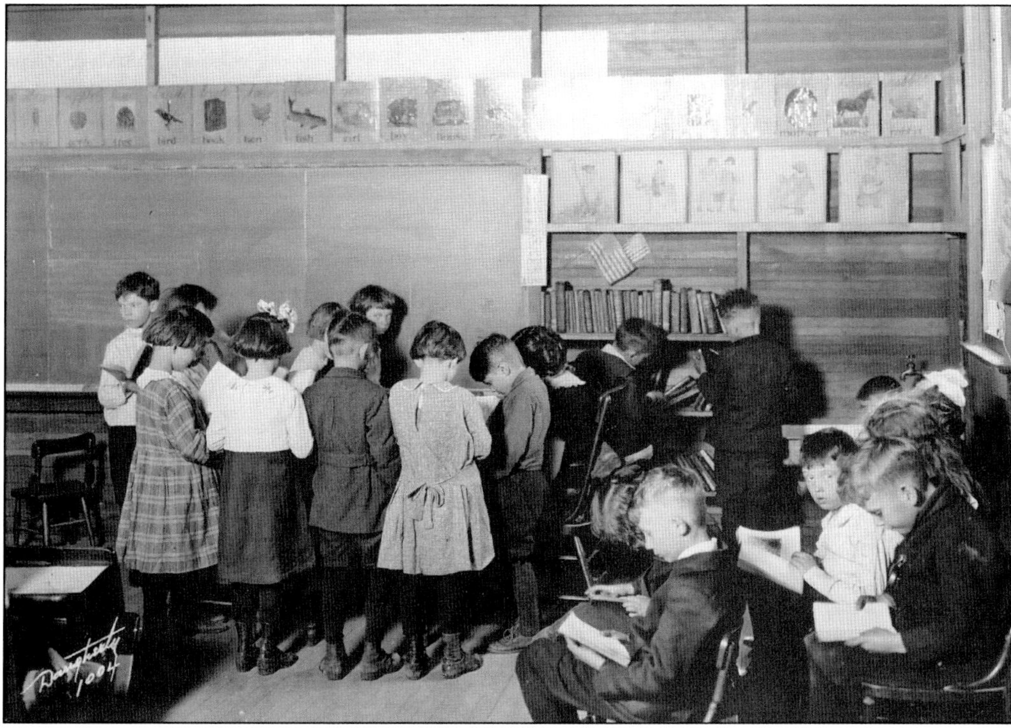

HORACE MANN PORTABLES, 1920S. With an increasing number of school-age children in the Steel City, the school board agreed to build portable classrooms for a few of the schools. Mann, Franklin, and later Roosevelt had portables in their early years. (Courtesy of CRA.)

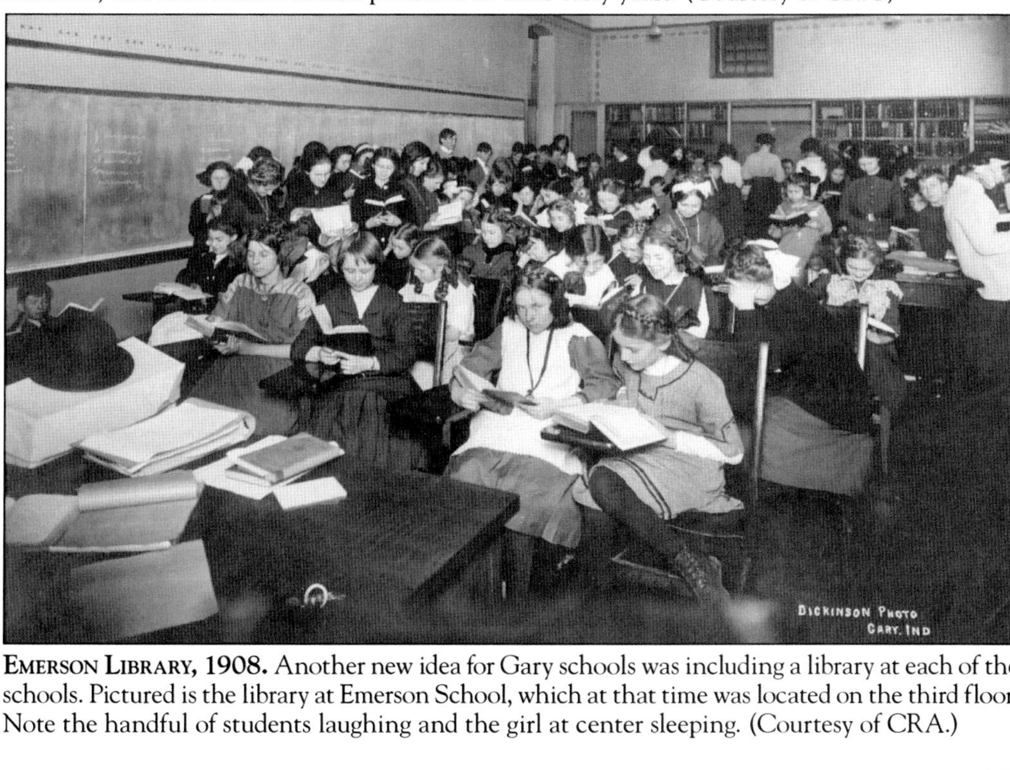

EMERSON LIBRARY, 1908. Another new idea for Gary schools was including a library at each of the schools. Pictured is the library at Emerson School, which at that time was located on the third floor. Note the handful of students laughing and the girl at center sleeping. (Courtesy of CRA.)

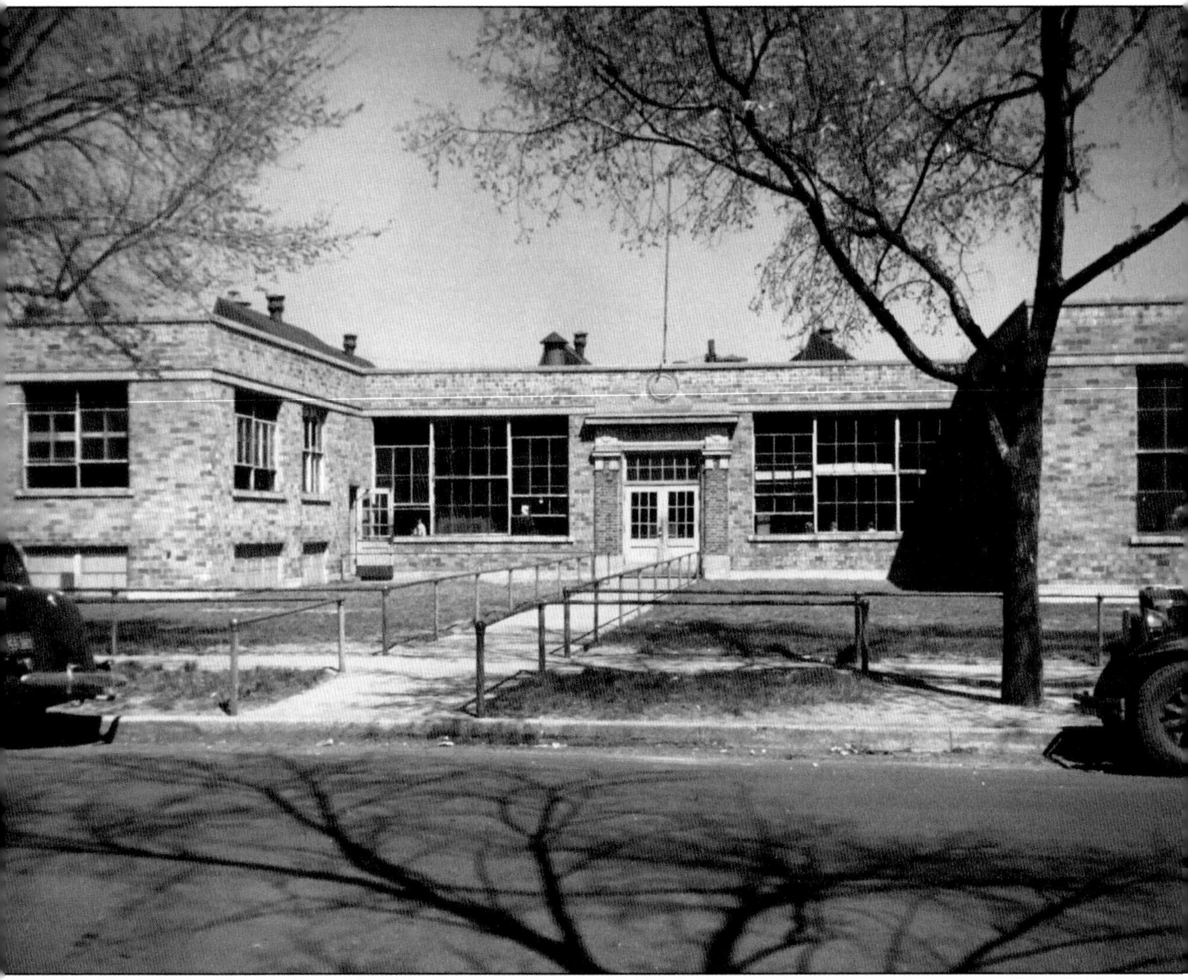

Emerson Shops, 1924. Vocational training was a very important part of the early Gary school curriculum as the main employer was US Steel. Most of the young men graduating from the city schools took a job at the big mill, as that was the best place for secure wages. (Courtesy of CRA.)

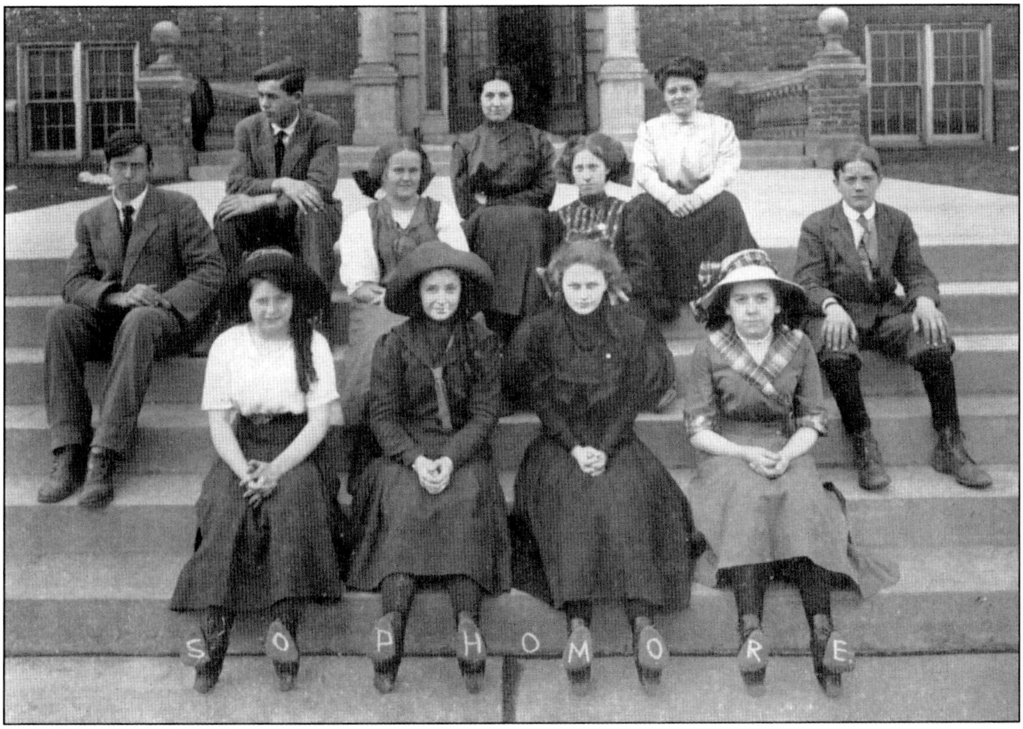

EMERSON SOPHOMORES, 1911. Pictured at Emerson's main entrance is the sophomore class in 1911. The clothing styles were quite different then. The unsinkable ship, the *Titanic*, would not make its maiden and last voyage for another year, and the Great War was still three years away. (Courtesy of CRA.)

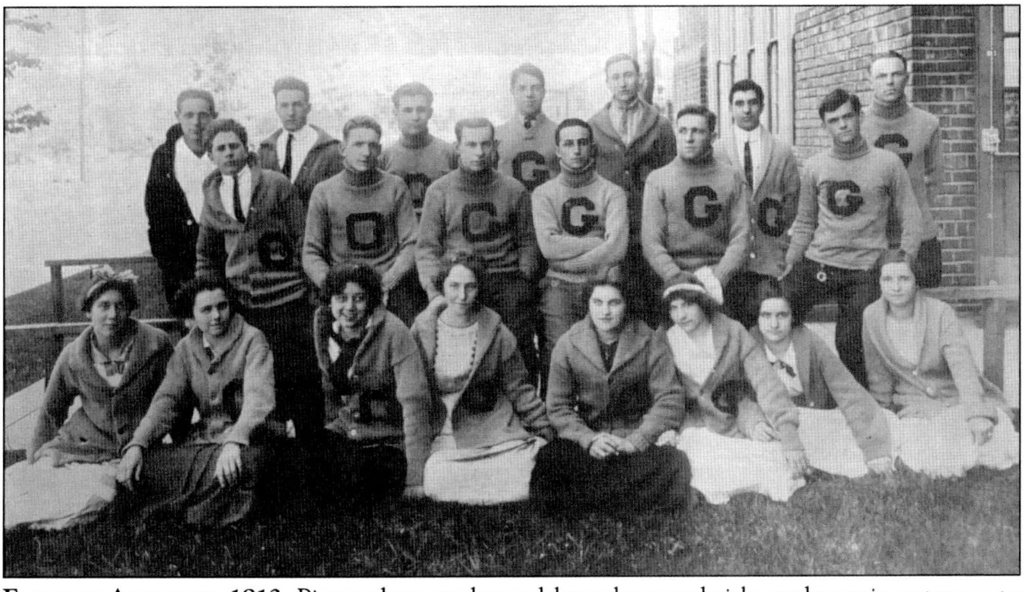

EMERSON ATHLETES, 1913. Pictured are student athletes, boys and girls, at the main entrance to Emerson. The reason for the letter G is that the school was known as Gary High School. When Froebel was built in 1913, it was part of Gary High School until 1917. After that, they went by Emerson or Froebel. (Courtesy of CRA.)

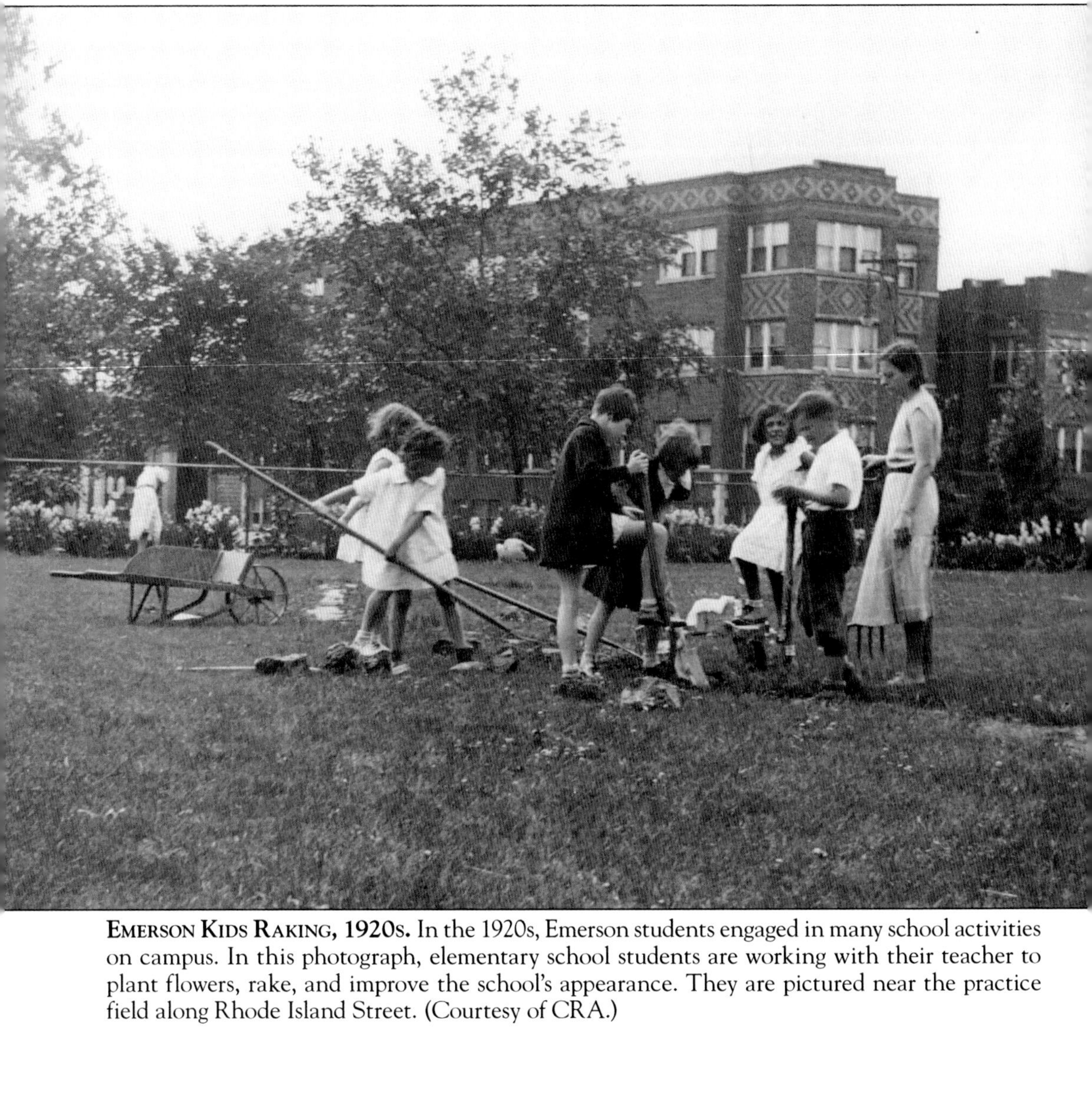

Emerson Kids Raking, 1920s. In the 1920s, Emerson students engaged in many school activities on campus. In this photograph, elementary school students are working with their teacher to plant flowers, rake, and improve the school's appearance. They are pictured near the practice field along Rhode Island Street. (Courtesy of CRA.)

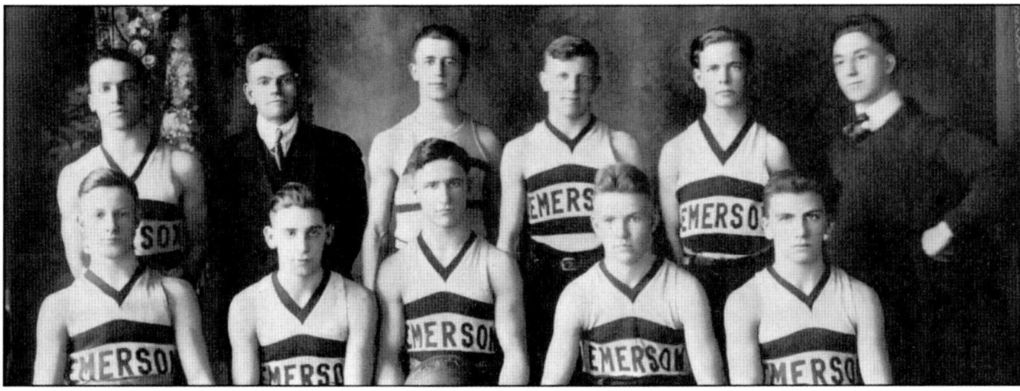

EMERSON 1917 BASKETBALL TEAM. Emerson made its only appearance in the Indiana State Basketball Final in 1917. In the championship, they faced a talented Lebanon team and lost 34-26. In the morning game, Emerson defeated Rochester 27-17. Lou Erickson was coach. (Courtesy of CRA.)

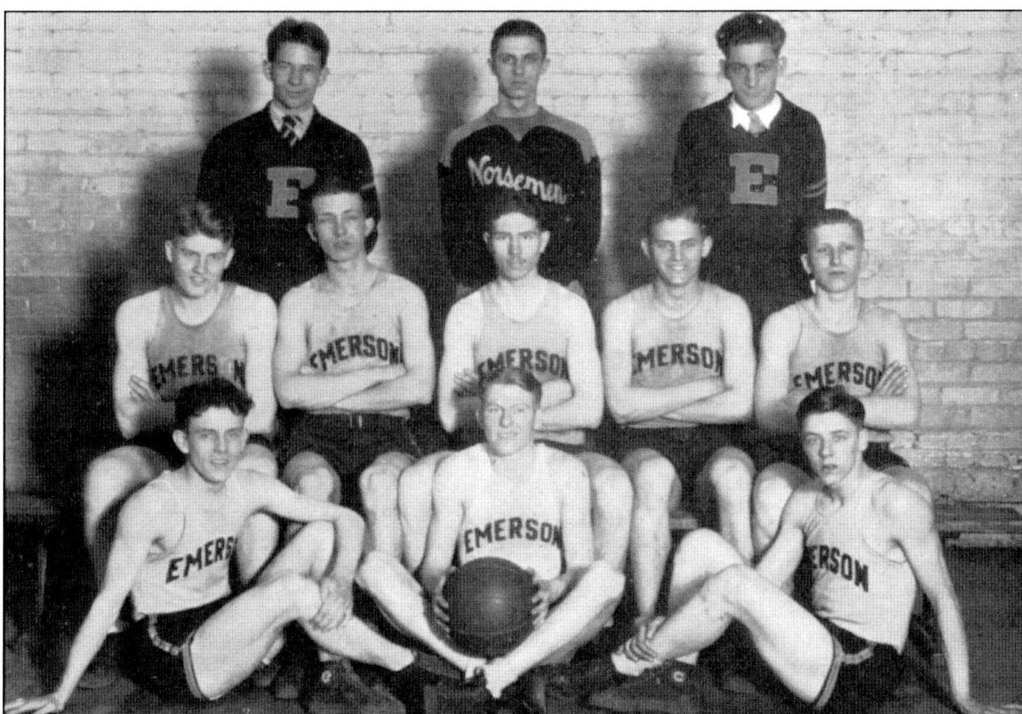

EMERSON 1927 SEMI-STATE CHAMPIONS. In 1927, Emerson's basketball championship hopes ended in the semi-state round as the team lost to an excellent Martinsville team led by John Wooden. Wooden went on to succeed as a player at Purdue, and later coached at the University of California, Los Angeles. (Courtesy of CRA.)

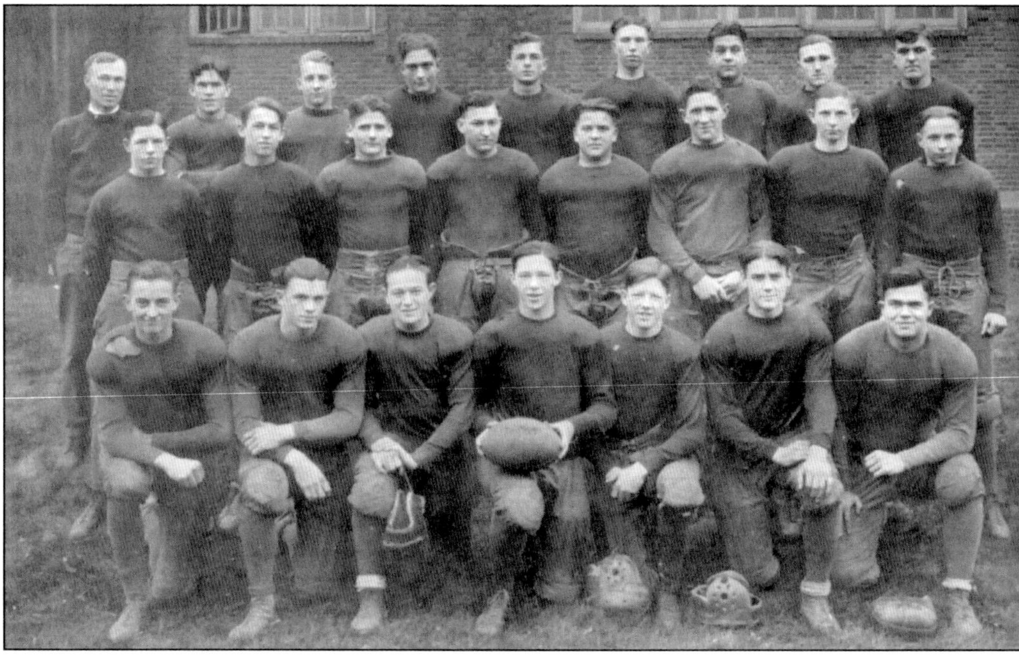

Emerson Football, 1925. Emerson High School had great success on the gridiron from its early years through the 1950s. From 1912 through 1930, the school won or shared eight state football titles. Pictured is the 1925 team, which won the school's third title and went undefeated. It was the 1924 team that earned the name Golden Tornado for its ability to score quickly and often. Except for the football team, all Emerson sports teams were referred to as the Norsemen. The young men pictured here look pretty tough—almost as if they just came off a shift at the mill. (Courtesy of CRA.)

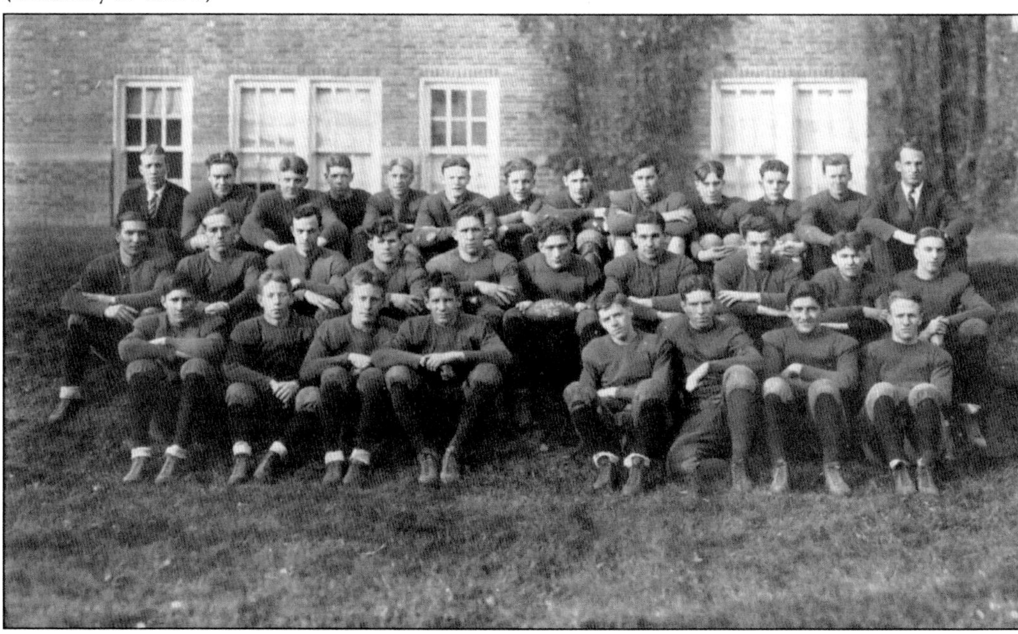

Emerson Football, 1926. The 1926 team won the state title and went undefeated. They are pictured on the south side of the school building with the game ball. (Courtesy of CRA.)

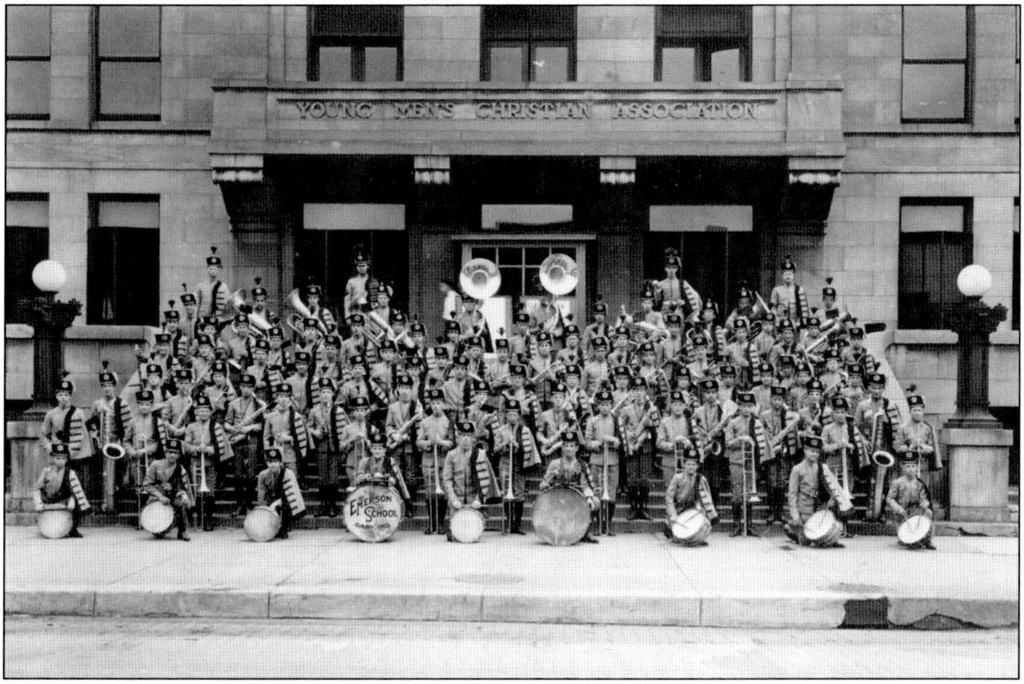

EMERSON BAND, 1925. Pictured in 1925 is the Emerson High School marching band in front of the YMCA building on West Fifth Avenue. Emerson's award-winning band traveled to a number of cities, including Washington, DC. (Courtesy of CRA.)

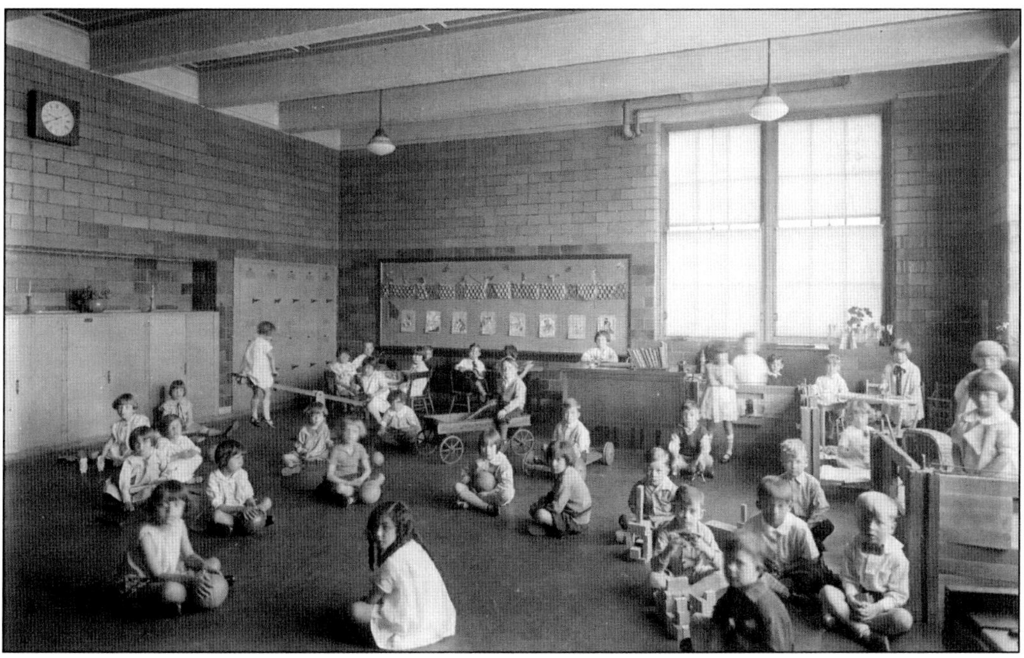

EMERSON ELEMENTARY, 1920s. In the late 1920s, a new building was completed just east of Emerson High School and served elementary-age children. The "little building," as some of the younger students called it, was for those from kindergarten through sixth grade. Pictured are kindergarten students around 1929. (Courtesy of CRA.)

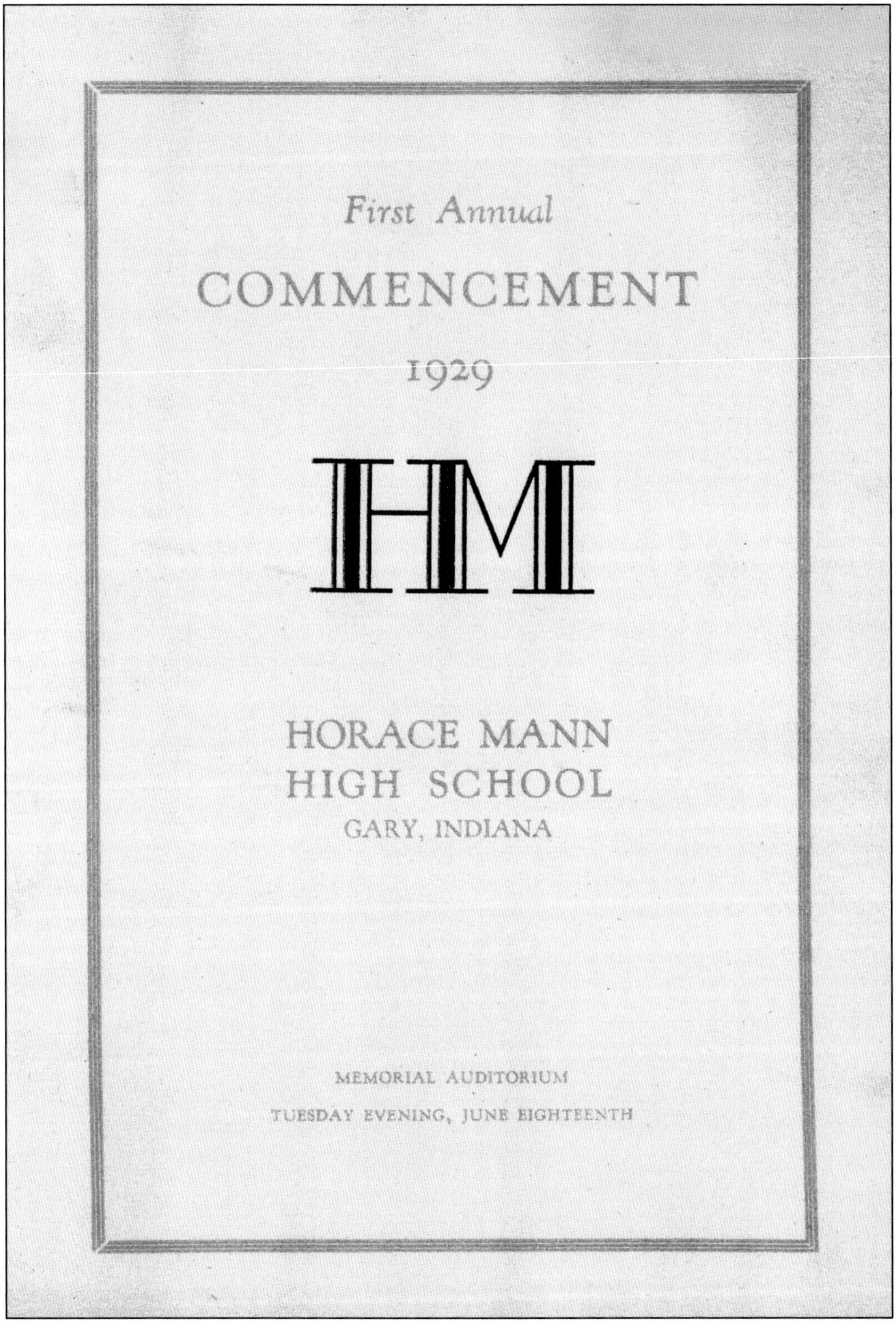

MANN COMMENCEMENT, 1929. On Tuesday June 18, 1929, Horace Mann held its first annual commencement in Memorial Auditorium at Seventh Avenue and Massachusetts Street. The graduates and their families had so much to look forward to that evening, but four months later, on October 29, Wall Street saw a huge drop. In time, the United States faced an economic disaster. (Courtesy of CRA.)

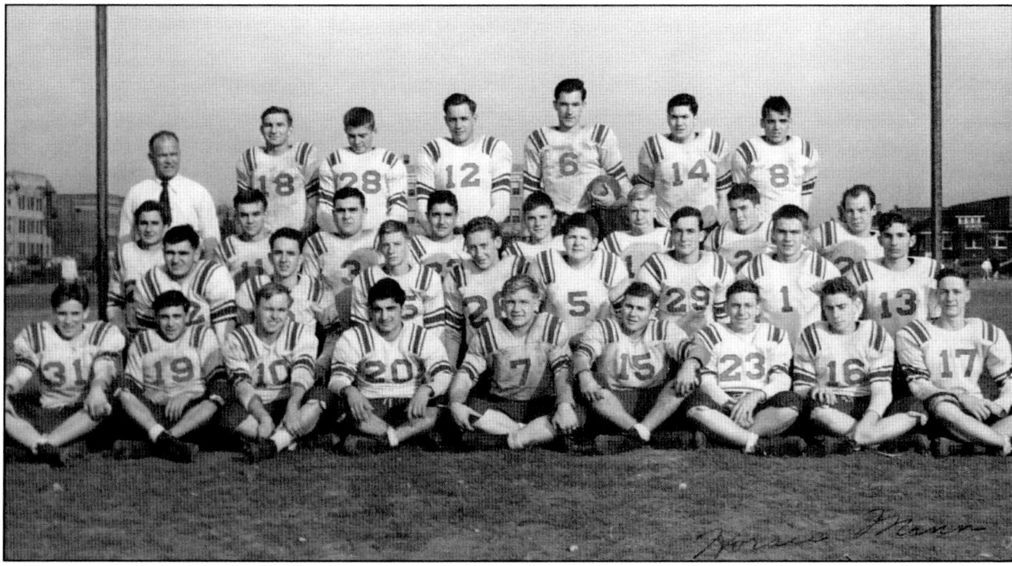

Horace Mann 1945 Team. Neither the Great Depression nor the two world wars suspended high school athletics at Gary schools. Budgets were limited, but the teams continued to play their schedules. Pictured is the 1945 Horace Mann High School football team on the practice field with coach Keith Crown. During the war, Horace Mann and other high school teams sent packages to their alumni who were fighting in Europe or the Pacific. Mann had a program called "For Our Boys" that sent care packages and letters to servicemen. Emerson had a similar project. (Courtesy of CRA.)

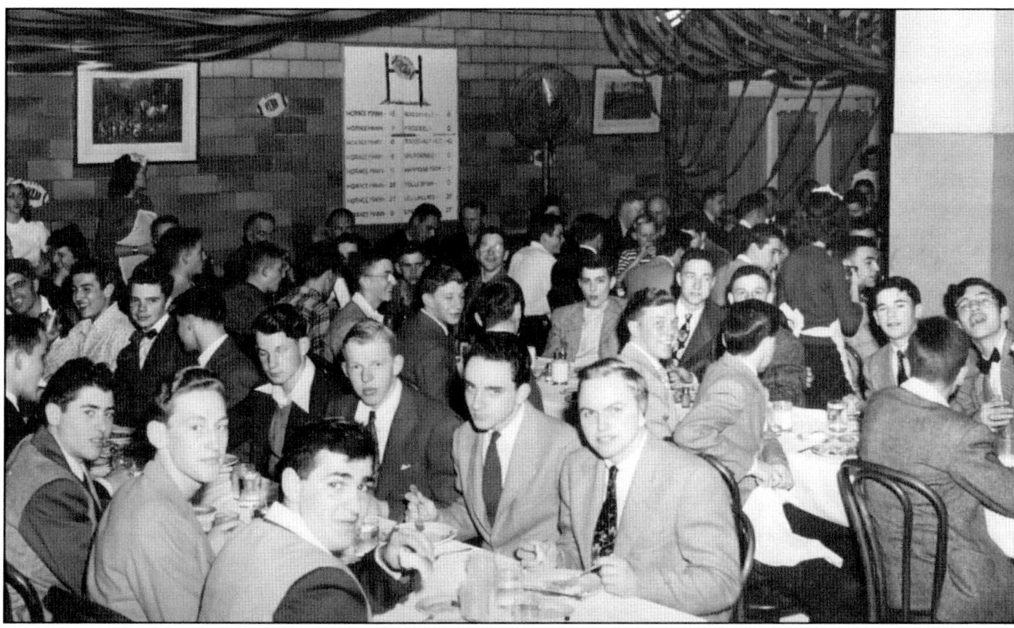

Mann Athletic Banquet, 1946. Athletic sports dinners were held at the close of the season to honor the players, coaches, and staff. Here, the Mann football team was honored for a winning season in 1945. That year, the team had a winning season of 4-3-1. In its history, Horace Mann won two state football titles and a number of conference championships. Emerson won five, Froebel two, and Lew Wallace and Gary Roosevelt one. (Courtesy of CRA.)

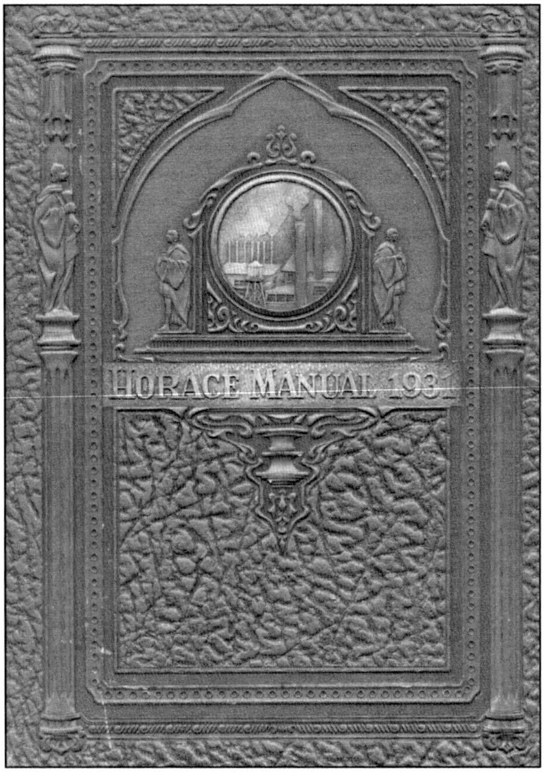

MANN YEARBOOK, 1931. In 1931, Mann continued to produce a high-quality, hardcover yearbook for its students, but as the Depression got worse, members of the yearbook staff knew a change was needed due to costs. (Courtesy of CRA.)

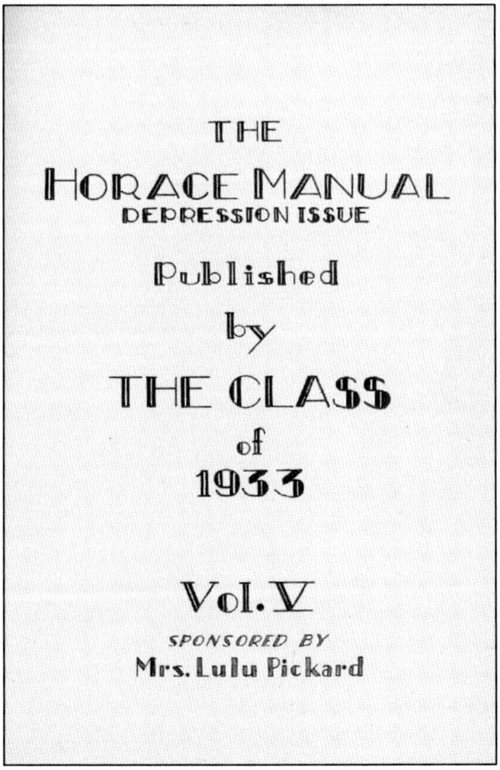
THE
HORACE MANUAL
DEPRESSION ISSUE
Published
by
THE CLASS
of
1933

Vol. V
SPONSORED BY
Mrs. Lulu Pickard

MANN YEARBOOK, 1933. The Mann yearbook had a soft cover this year because the school and its students were struggling due to the declining economy. Most families could not afford the yearbooks. (Courtesy of CRA.)

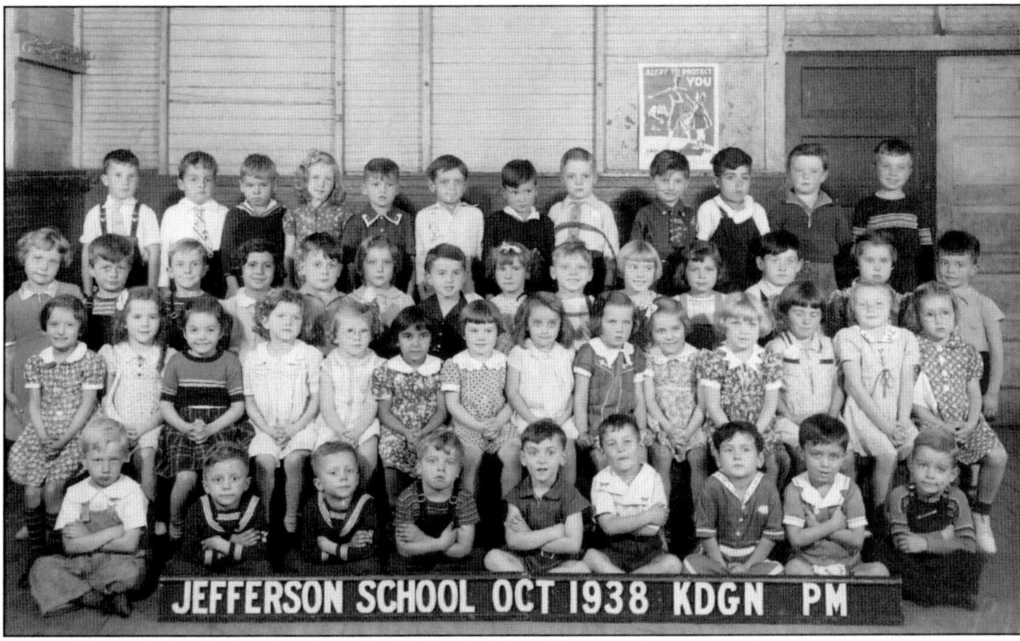

JEFFERSON SCHOOL KINDERGARTEN, 1938. Though the nation's economy was improving, families still had to make tough decisions concerning their personal finances. Secondhand clothes were common in the 1930s for most families in Gary. Many teenagers often quit school and went to work to help their parents. These youngsters were lucky to be in school. (Courtesy of CRA.)

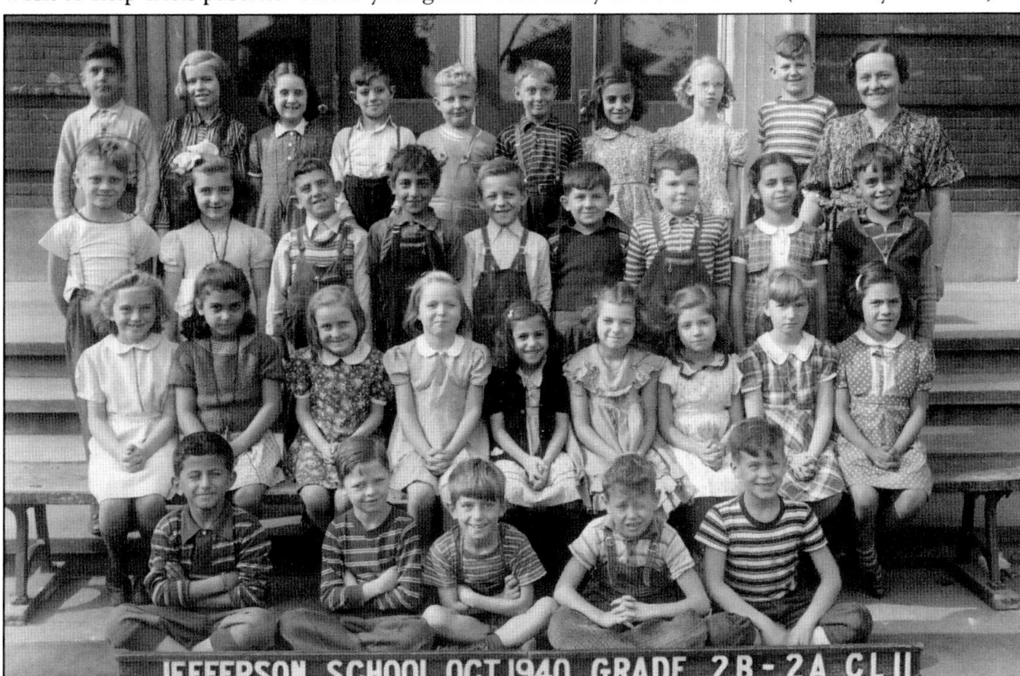

JEFFERSON SCHOOL SECOND GRADE, 1940. Pictured is the second-grade class of Jefferson School in 1940. Though the war raged across most of Europe, these youngsters were fortunate to still be part of a peacetime, neutral America. (Courtesy of CRA.)

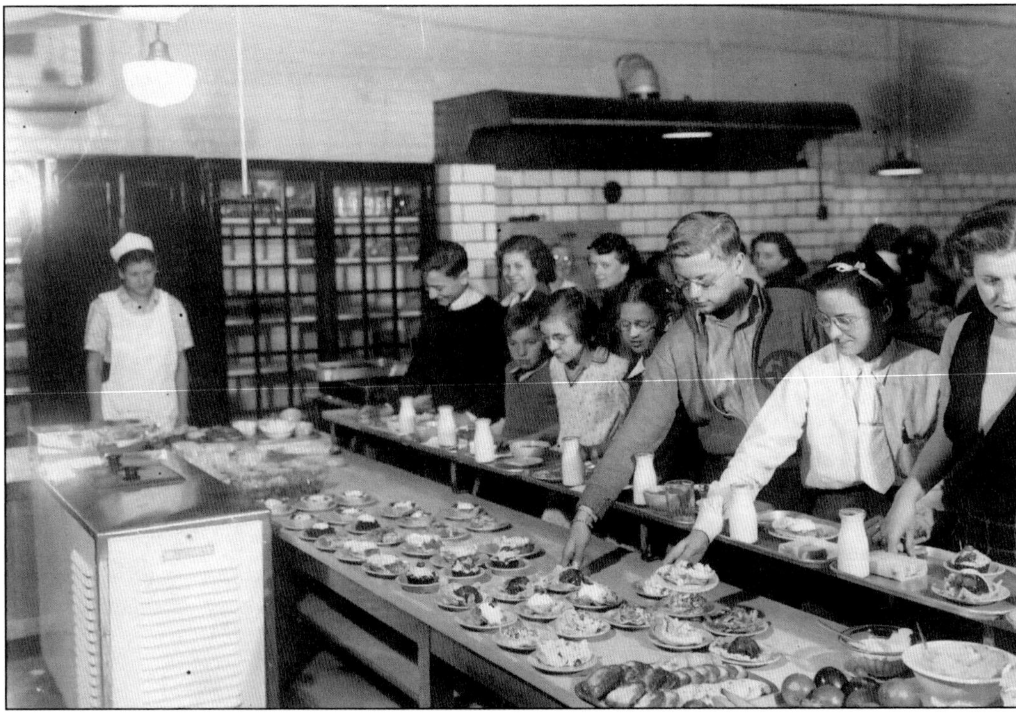

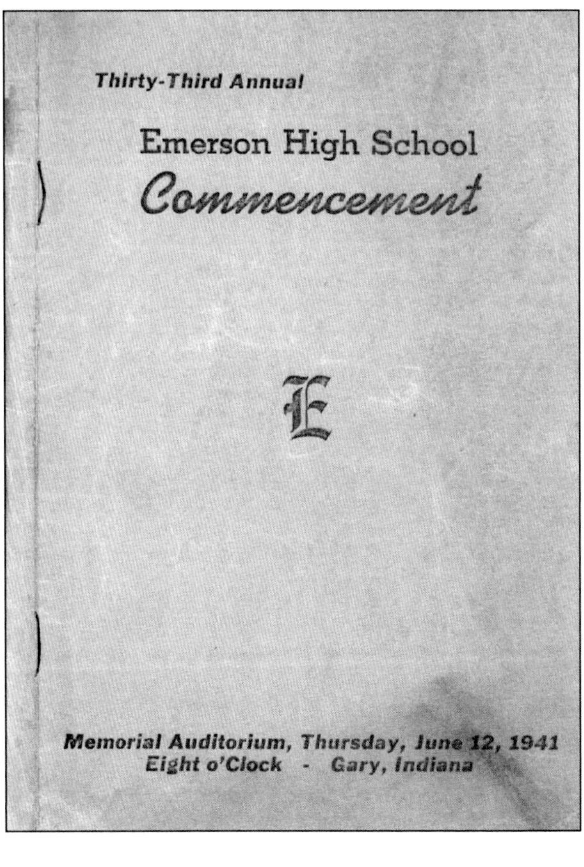

Thirty-Third Annual

Emerson High School

Commencement

E

Memorial Auditorium, Thursday, June 12, 1941
Eight o'Clock - Gary, Indiana

EMERSON CAFETERIA, 1930s. Part of Franklin D. Roosevelt's New Deal policies included money for lunch programs in the nation's schools. Many children faced hunger, especially in rural areas. By the late 1930s, most schools qualified for one meal for students. Pictured is the first-floor cafeteria at Emerson High School. (Courtesy of CRA.)

EMERSON COMMENCEMENT PROGRAM, 1941. In 1941, Emerson held its commencement ceremonies at Memorial Auditorium at Seventh Avenue and Massachusetts Street. In December of that year, Japan bombed Pearl Harbor, and the next day, war was declared. Most of the young men graduating that evening would be in the military during the conflict. All would know a family member, friend, or neighbor who served. (Courtesy of CRA.)

Emerson Reserve Officers' Training Corps, 1943. The main theme of this page from the 1943 Emerson yearbook was patriotism. This photograph was actually taken the previous year in 1942 to be used for the 1943 book. (Courtesy of CRA.)

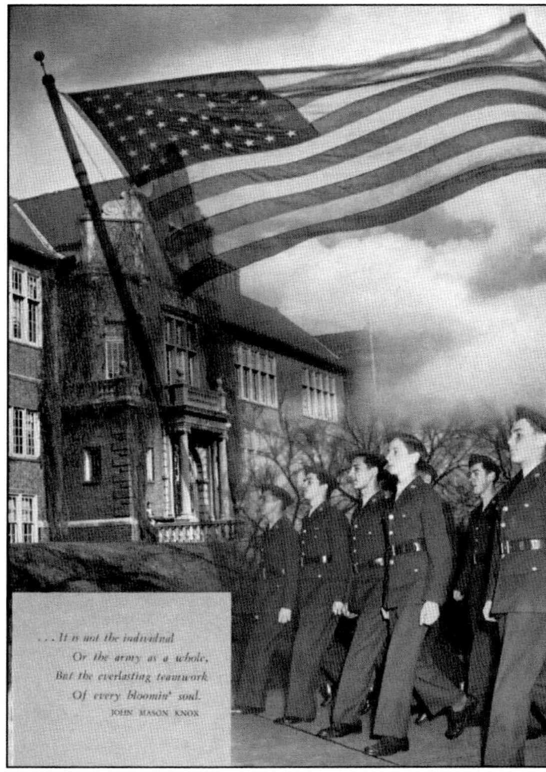

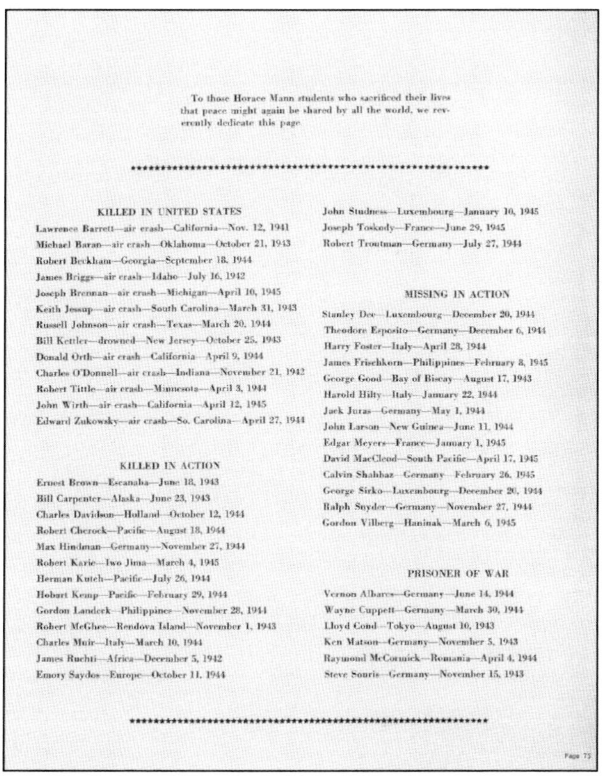

To those Horace Mann students who sacrificed their lives that peace might again be shared by all the world, we reverently dedicate this page.

****

KILLED IN UNITED STATES

Lawrence Barrett—air crash—California—Nov. 12, 1941
Michael Baran—air crash—Oklahoma—October 21, 1943
Robert Beckham—Georgia—September 18, 1944
James Briggs—air crash—Idaho—July 16, 1942
Joseph Brennan—air crash—Michigan—April 10, 1945
Keith Jessup—air crash—South Carolina—March 31, 1943
Russell Johnson—air crash—Texas—March 20, 1944
Bill Kettler—drowned—New Jersey—October 25, 1943
Donald Orth—air crash—California—April 9, 1944
Charles O'Donnell—air crash—Indiana—November 21, 1942
Robert Tittle—air crash—Minnesota—April 3, 1944
John Wirth—air crash—California—April 12, 1945
Edward Zukowsky—air crash—So. Carolina—April 27, 1944

KILLED IN ACTION

Ernest Brown—Escanaba—June 18, 1943
Bill Carpenter—Alaska—June 23, 1943
Charles Davidson—Holland—October 12, 1944
Robert Cherock—Pacific—August 18, 1944
Max Hindman—Germany—November 27, 1944
Robert Karie—Iwo Jima—March 4, 1945
Herman Kutch—Pacific—July 26, 1944
Hobart Kemp—Pacific—February 29, 1944
Gordon Landeck—Philippines—November 28, 1944
Robert McGhee—Rendova Island—November 1, 1943
Charles Muir—Italy—March 10, 1944
James Ruchti—Africa—December 5, 1942
Emory Saydos—Europe—October 11, 1944
John Studness—Luxembourg—January 10, 1945
Joseph Toskody—France—June 29, 1945
Robert Troutman—Germany—July 27, 1944

MISSING IN ACTION

Stanley Dee—Luxembourg—December 20, 1944
Theodore Esposito—Germany—December 6, 1944
Harry Foster—Italy—April 28, 1944
James Frischkorn—Philippines—February 8, 1945
George Good—Bay of Biscay—August 17, 1943
Harold Hilty—Italy—January 22, 1944
Jack Juras—Germany—May 1, 1944
John Larson—New Guinea—June 11, 1944
Edgar Meyers—France—January 1, 1945
David MacCleod—South Pacific—April 17, 1945
Calvin Shahbaz—Germany—February 26, 1945
George Sirko—Luxembourg—December 20, 1944
Ralph Snyder—Germany—November 27, 1944
Gordon Vilberg—Haninak—March 6, 1945

PRISONER OF WAR

Vernon Albares—Germany—June 14, 1944
Wayne Cuppett—Germany—March 30, 1944
Lloyd Cond—Tokyo—August 10, 1943
Ken Matson—Germany—November 5, 1943
Raymond McCormick—Romania—April 4, 1944
Steve Souris—Germany—November 15, 1943

****

Page 75

Horace Mann War Losses, 1945. In 1945, Horace Mann High School decided to honor all of its graduates who served during the war. Listed here are those missing in action, taken prisoners of war, and killed in action, the ultimate sacrifice for the nation's freedom. (Courtesy of CRA.)

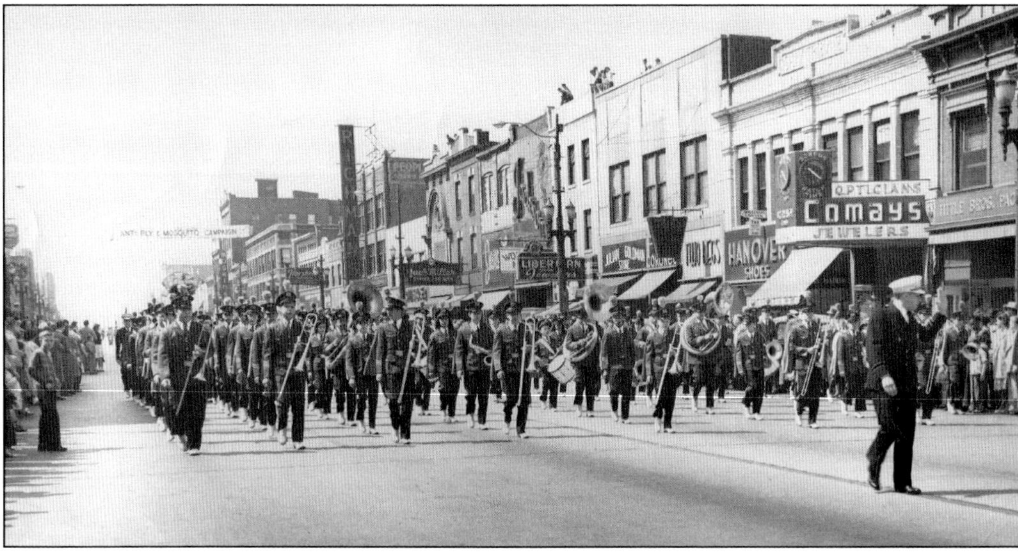

BUG ABATEMENT PARADE, 1953. In 1953, the City of Gary went all out to promote the bug and mosquito abatement program by reaching out to schools, employers, and residents. A parade was held to kick off the summer's events. Pictured is the Froebel High School band marching up the 600 block of Broadway. (Courtesy of CRA.)

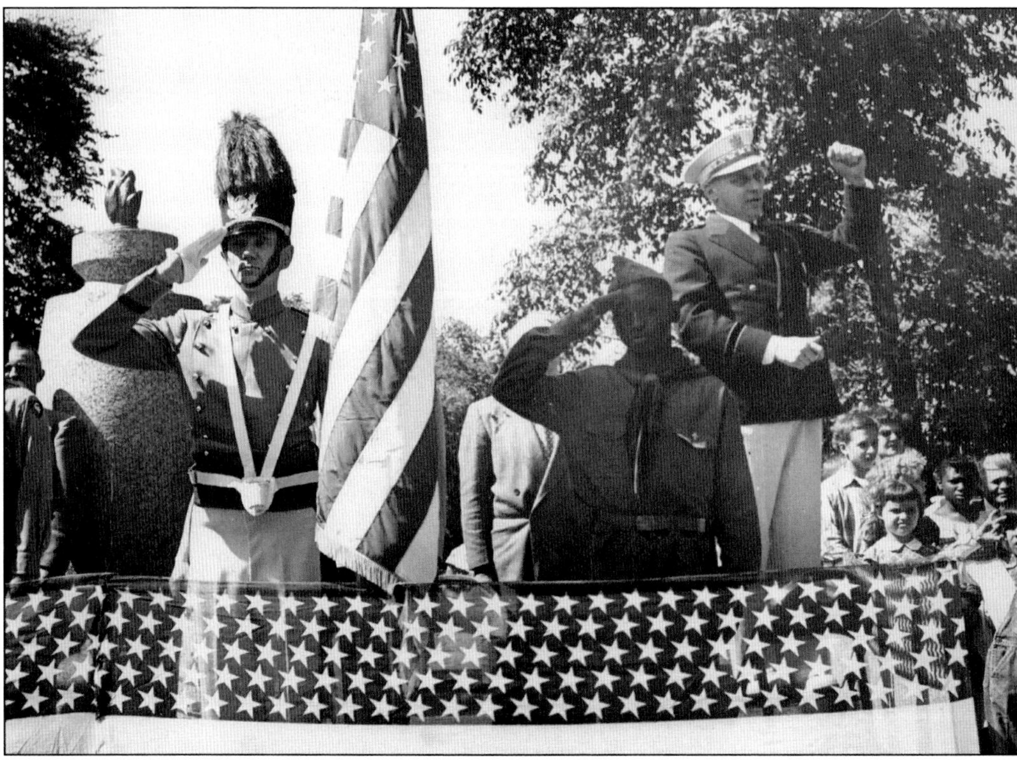

KEN RESUR LEADING BANDS AT BUFFINGTON PARK, 1953. Ken Resur, longtime band director of Froebel, later Horace Mann High School, is shown leading the city bands in Buffington Park at Sixth Avenue and Connecticut Street. The 1953 Memorial Day event was held to honor the nation's war dead. (Courtesy of CRA.)

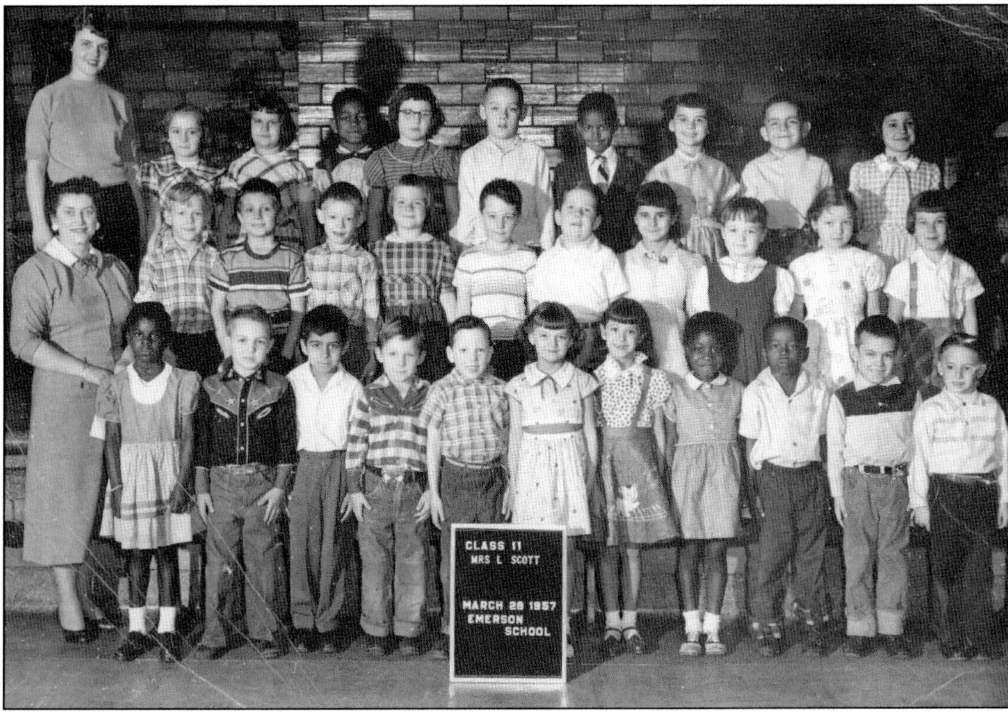

EMERSON FIRST GRADE, 1957. In March 1957, the students of Mrs. Scott's Class No. 11 dressed nicely for their class picture, taken at Emerson Elementary School. Perhaps a few took their time to smile and pose in order to stay out of class a little longer. Most of the group went through Emerson together all the way to high school graduation. (Courtesy of Judy and Dave Trathen.)

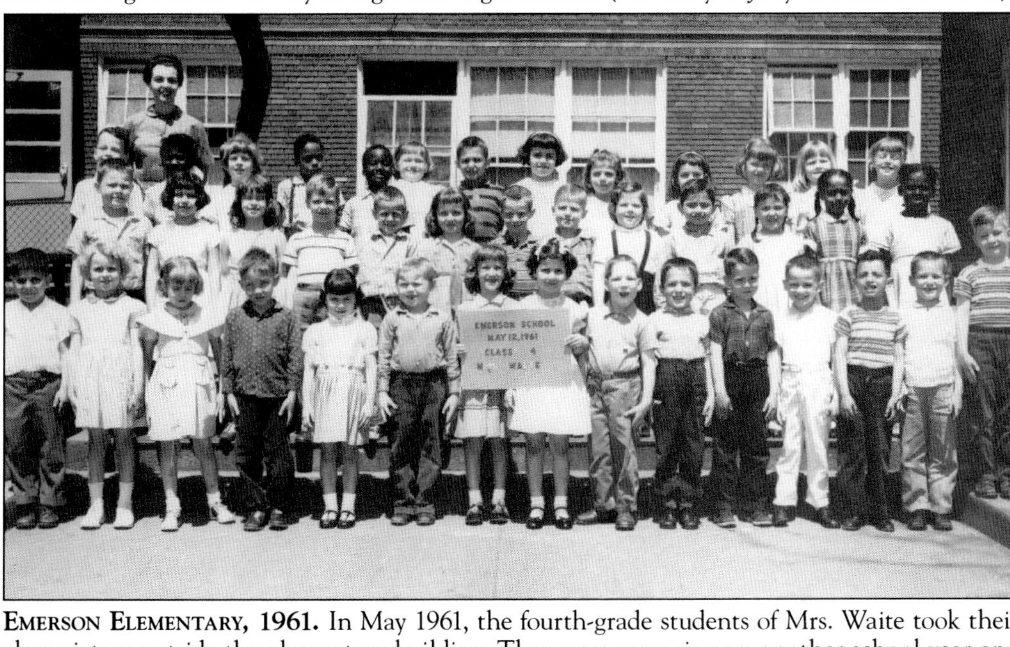

EMERSON ELEMENTARY, 1961. In May 1961, the fourth-grade students of Mrs. Waite took their class picture outside the elementary building. They were wrapping up another school year and were probably looking forward to the upcoming summer break, just like the teachers. (Courtesy of CRA.)

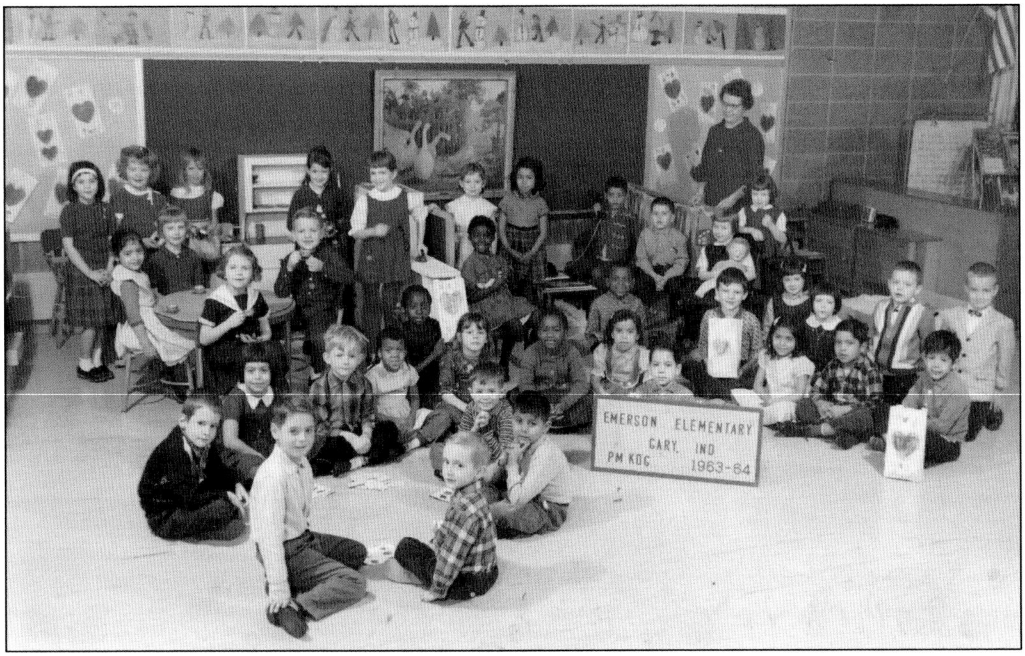

Mrs. Wilson's Class. In the late 1950s, on the northwest corner of Seventh Avenue and Rhode Island Street, a new school building, named after Emerson's first principal, E.A. Spaulding, was built on the Emerson campus for students from kindergarten through fourth grade. Pictured is Mrs. Wilson's kindergarten class in 1963. (Courtesy of Diane Trafny.)

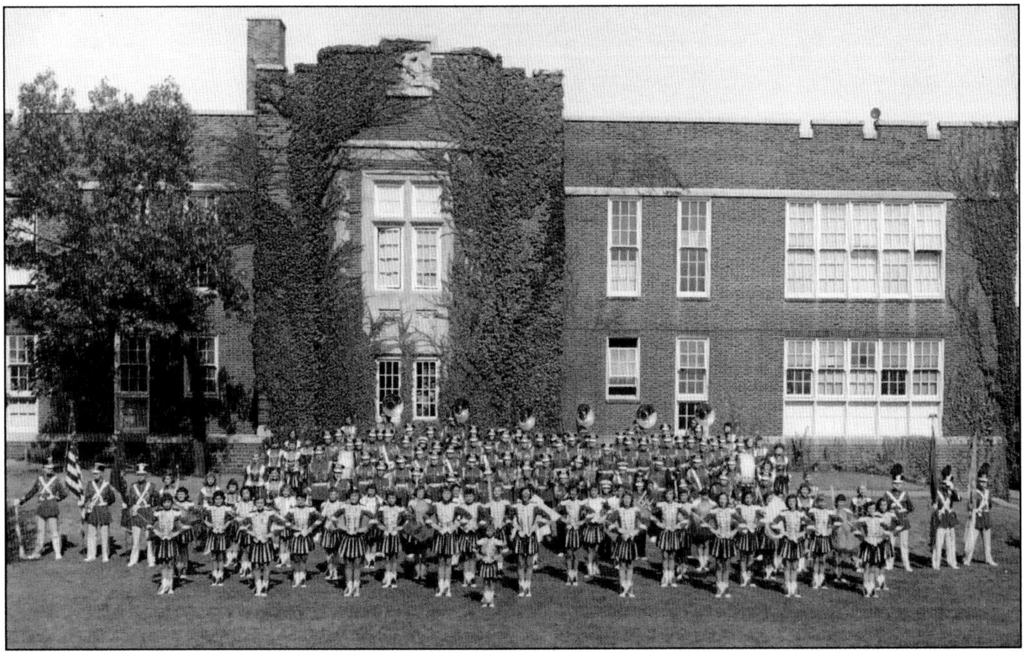

Horace Mann Band, 1960s. Pictured is the 1960 Horace Mann band in front of the south side of the school. The band was directed for many years by Ken Resur and his wife, Anne, who was in charge of the majorettes. Under his leadership, Mann's band became an award-winning band in the Calumet Region. (Courtesy of CRA.)

EMERSON TURNABOUT, 1966. In December 1966, Linda Swallow (Benka) and John Trafny went to the Emerson High School Christmas turnabout that was held at the Hotel Gary's Chrystal Ballroom downtown. For this event, the young ladies had to ask the young men. (Courtesy of John Trafny.)

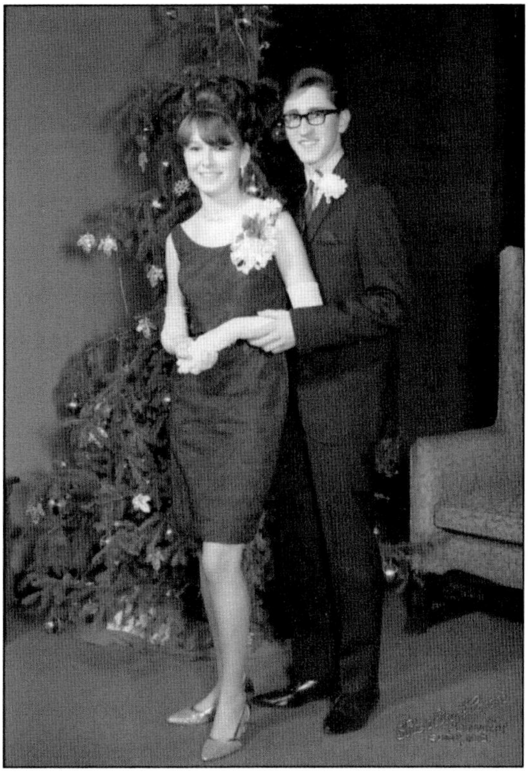

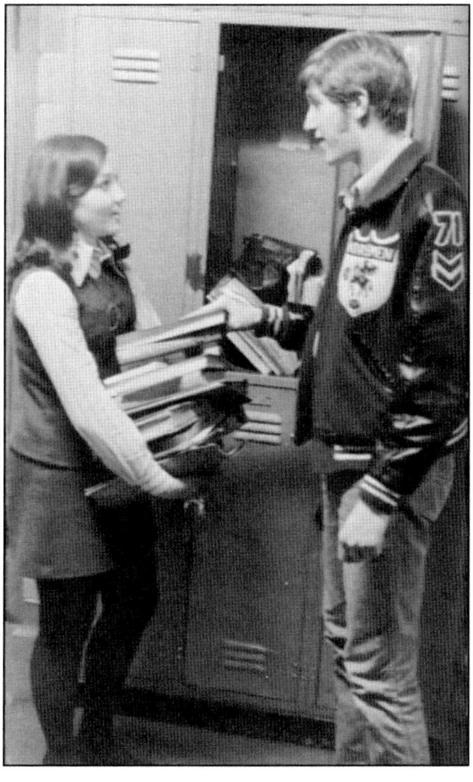

HORACE MANN TWIRP WEEK, 1969. Students in the Gary high schools had traditions that probably would not be allowed today. At Emerson, the Girl's Athletic Association (GAA) freshmen girls had to follow initiation rules to join. They had to dress like slobs during Scum Week and go through air raids during that time. At Mann, the girls had to attend to daily care for the boys. They had to carry books, hold doors, tie their shoes, and just look after the guys, who probably were accustomed to mommy attending to them at home. (Courtesy of CRA.)

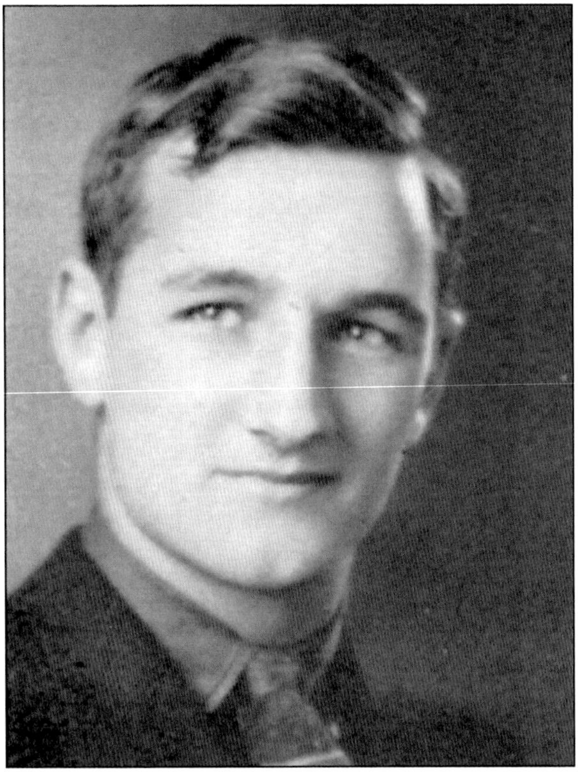

Tom Harmon, 1937. Tom Harmon was an all-star athlete at Horace Mann and later at the University of Michigan, where he won the Heisman Trophy in 1940. During World War II, he served as a captain in the US Army Air Corps. He did play for a few years with the Rams but went on to sports broadcasting. His son Mark is the star of the CBS television show *NCIS*. (Courtesy of CRA.)

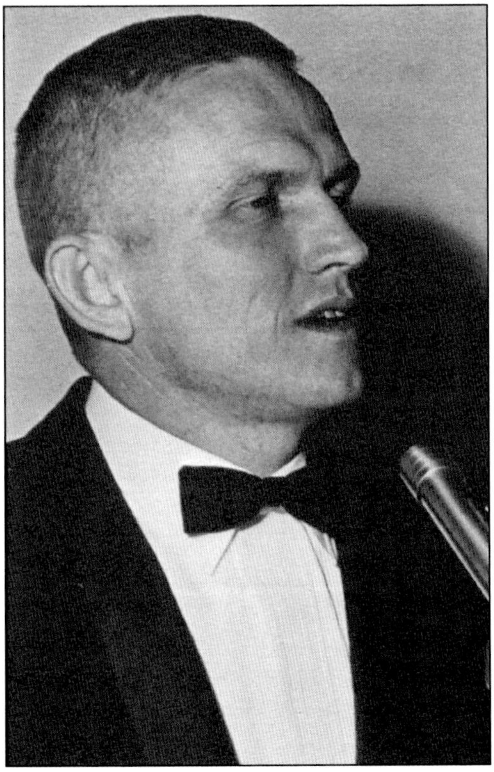

Frank Borman, 1967. Astronaut Frank Borman was born in the Horace Mann neighborhood, but his family moved to Arizona when he was a teenager. Borman graduated from West Point and became a fighter pilot in the Air Force. In the 1950s, he was part of the Gemini Program, and in 1968, he and his crew on Apollo 8 orbited the moon on the evening of December 24, Christmas Eve. The event was televised and helped close the challenging year of 1968 on a positive note. Borman later became president of Eastern Airlines. (Courtesy of CRA.)

Karl Malden. 1931 Emerson graduate Karl Malden went on to great success in Hollywood in movies and television. Malden Sekulovic grew up a block from the high school, studied drama at DePaul University, and worked a few years at the open hearth department at US Steel's Gary Works. During World War II, he serve as a noncommissioned officer in the US Army Air Corps. Some of his movies were *Patton*, *Birdman of Alcatraz*, and *On the Waterfront*. On television, he played a lead in *The Streets of San Francisco* on ABC and *Skag* on NBC. (Courtesy of CRA.)

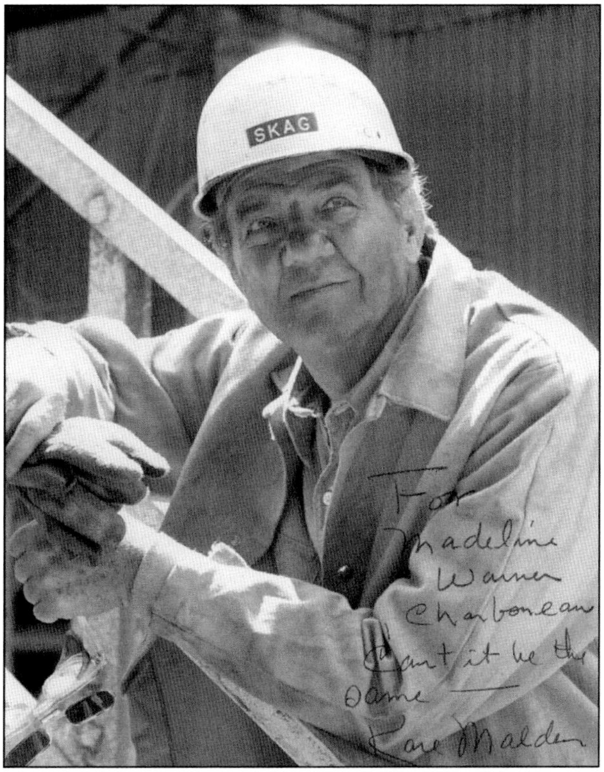

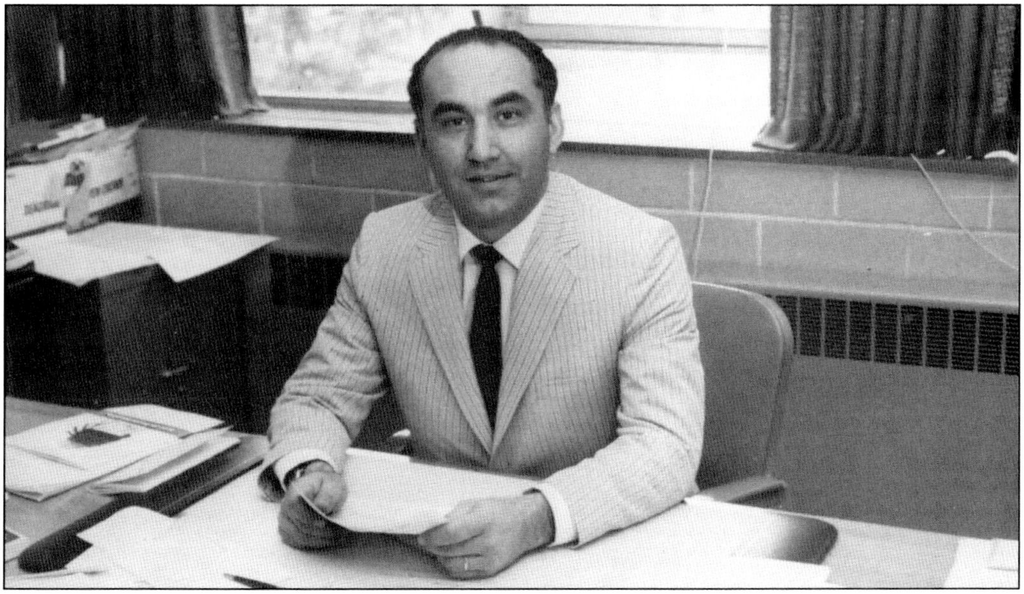

Christ Christoff, 1971. Christ Christoff, longtime educator in Gary schools, is pictured here in 1970. The Froebel graduate taught at Emerson High School in the 1960s and then went on to be principal of his alma mater and then assistant principal at West Side High School. In the 1970s, he served as principal at Lew Wallace High School in the Glen Park neighborhood. After his retirement, he continued to be greatly respected by the Emerson High School alumni. (Courtesy of CRA.)

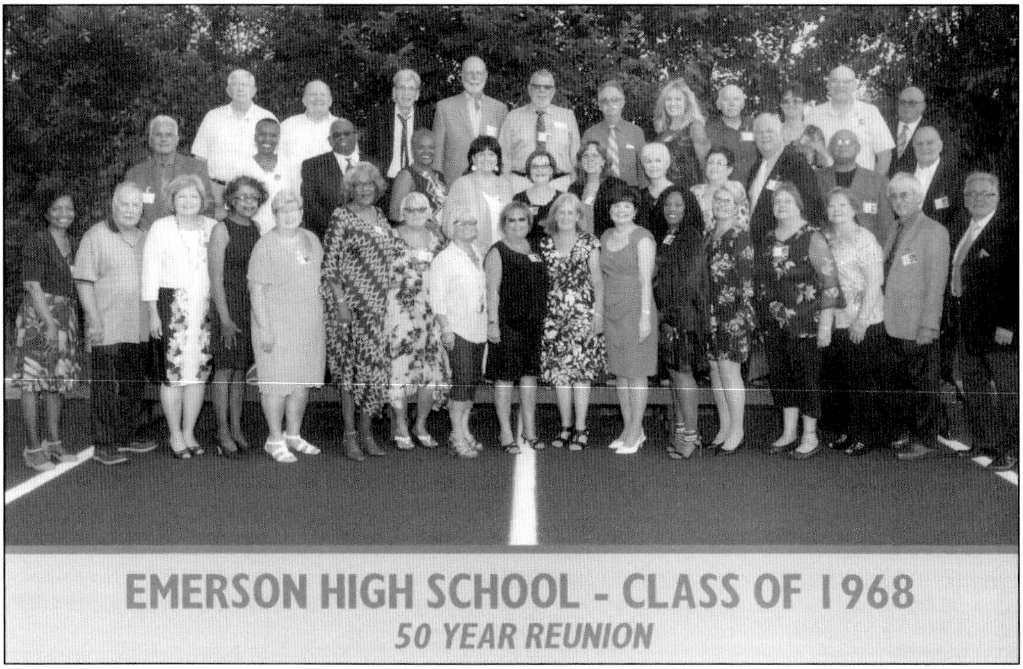

EMERSON CLASS OF 1968, 50TH REUNION. In August 2018, the Emerson class of 1968 met for their 50th class reunion. When they graduated, the nation was divided over the war in Vietnam, issues of race, and changing values and feelings of the nation and world that they knew. But like the previous generations that faced their own challenges, the class of 1968 endured and pursued their educations or job skills, got jobs, raised families, and became productive citizens. Some of the members of that class are pictured here. (Courtesy of John Trafny.)

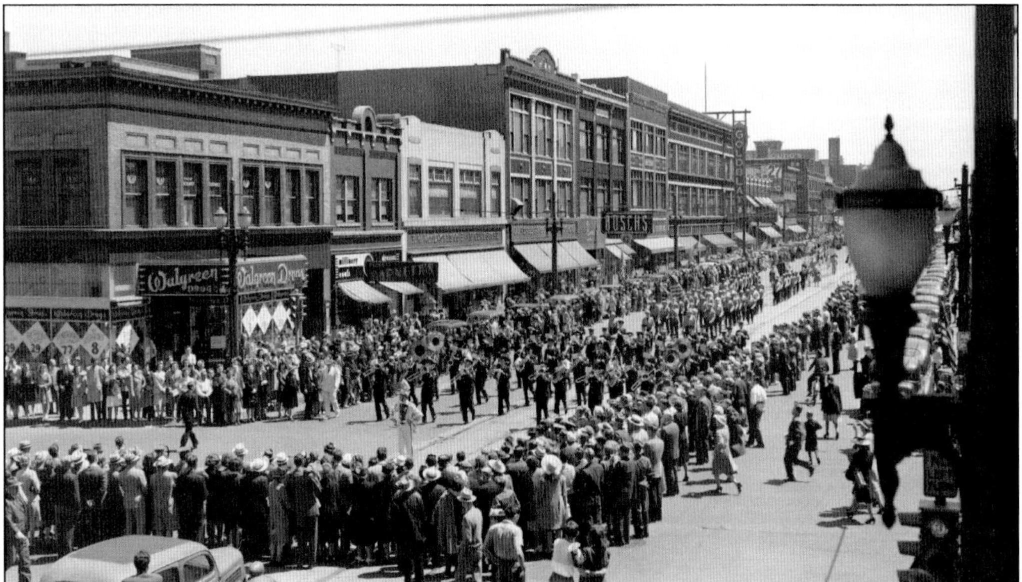

A FINAL LOOK DOWNTOWN. Pictured is the Memorial Day parade of 1948, looking south on Broadway from Sixth Avenue. Hopefully, this book brought back fond memories for the reader, their families, and friends of the city of Gary and its great downtown. (Courtesy of CRA.)

Bibliography

Bertino, Dominic V. *In the Presence of Angels: A History of the Diocese of Gary.* Strasbourg, France: Éditions du Signe, 2006.

Cohen, Ronald D. and James B. Lane. *Gary: A Pictorial History.* Virginia Beach, VA: Donning Company Publishers, 1983 and 2003.

Lane, James B. *City of the Century: A History of Gary, Indiana.* Bloomington, IN: Indiana University Press, 1978.

Moore, Powell A. *The Calumet Region: Indiana's Last Frontier.* Indiana Historical Bureau, 1959.

Svengalis, Kendall. *Gary Indiana: A Centennial Celebration.* Guilford, CT: Duneland Press, 2006.

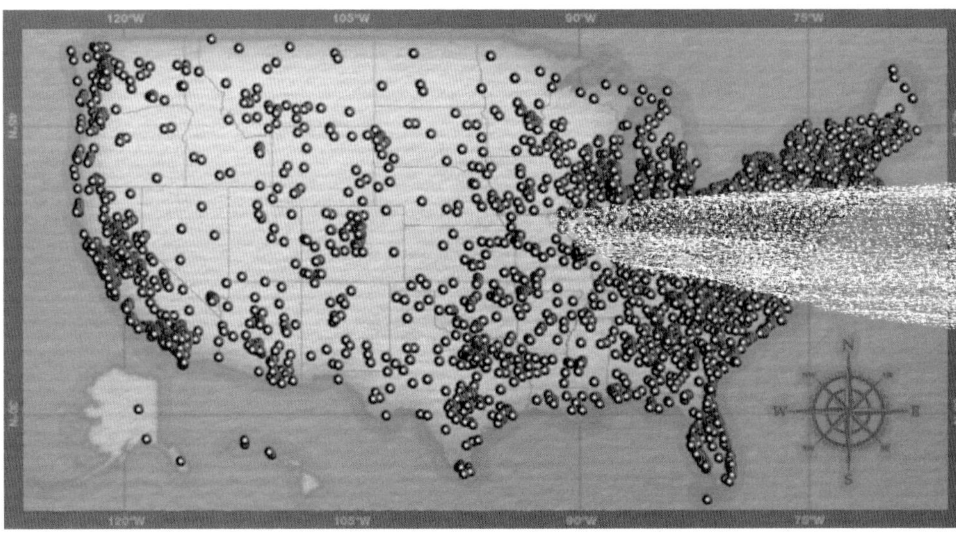